Driving IT Innovation

A Roadmap for CIOs to Reinvent the Future

Heather A. Smith
James D. McKeen

Prospect Press

Founded in 2014, Prospect Press serves the discipline of Information Systems by publishing innovative textbooks across the academic curriculum as well as select trade titles for IT professionals. Based in Burlington, Vermont, Prospect Press distributes worldwide. We welcome new authors to send proposals or inquiries to Beth.Golub@ProspectPressVT.com.

Editor: Beth Lang Golub
Assistant Editor: Tom Anderson-Monterosso
Production Management: Rachel Paul
Cover Design: Annie Clark

eBook ISBN 978-1-943153-63-3

Paperback ISBN 978-1-943153-64-0

For more information, visit DrivingITInnovation.com.

Contents

Acknowledgments

The work contained in this book is based on numerous meetings with many senior IT executives and managers. We would like to acknowledge our indebtedness to the following individuals who willingly shared their insights based on their experiences "earned the hard way":

Michele Baughman, Nastaran Bisheban, Peter Borden, Martin Boyer, Vince Campanelli, Lindsay Cartwright, Ryan Coleman, Dennis Dalmas, Randall Davis, Michael East, Michael Eubanks, Dave Forest, Gene Genin, Doug Gerhart, Erin Golding, Alaisdar Graham, Vineet Gupta, Chris Guzzo, Chris Harvison, Richard Hayward, Scott Ion, Jim Irich, Alnoor Jiwani, Grace Kennedy, Sherman Lam, Ron McKelvie, Celso Mello, Sean O'Farrell, Kelly Kanellakis, Kashif Parvaiz, Brian Patton, Klaus Peltsch, Robert Power, Pat Ressa, Alexey Rikhva, Katia Saenko, Tony

Saini, Adina Saposnik, Ashish Saxena, Linda Siksna, Raymond Smolskis, Kartik Subramani, Bruce Thompson, Joseph Trohak, and Julia Zhu.

Heather A. Smith
James D. McKeen
October 2018

About the Authors

Heather A. Smith has been named the most-published researcher on IT management issues in two successive studies (2006, 2009). A senior research associate with Smith School of Business at Queen's University, she is the author of eight books, the most recent being *IT Strategy and Innovation*, 4th ed. (Prospect Press, 2018). She is also a senior research associate with the American Society for Information Management's Advanced Practices Council. A former senior IT manager, she is co-director of the IT Management Forum and the CIO Brief, which facilitate interorganizational learning among senior IT executives. Heather consults and collaborates with organizations worldwide.

James D. McKeen is Professor Emeritus at the Smith School of Business at Queen's University and Senior Vice President and Chief Technology Officer at Empire Life Insurance. He has worked in the IT field for many years as a practitioner, researcher, and consultant. In 2011, he was named the "IT Educator of the Year" by *ComputerWorld Canada*. Jim has taught at universities in Canada, New Zealand, United Kingdom, France, Germany, Ukraine, and the United States. His research is widely published in a number of leading journals, and he is the coauthor (with Heather Smith) of eight books on IT strategy and management. Their 2015 book *IT Strategy: Issues and Practices* was the best-selling IT Strategy textbook in the United States.

Introduction

nnovation is (and has always been) acting on ideas that create value. What is new is the speed at which changes are taking place. And what's behind this speed is our ability to harness technology to drive "better, faster, cheaper." In the time it took to ask "What is Uber?," the taxi industry was in decline. Long before Waze had entered their lexicon, city planners were struggling to explain sudden traffic flow changes in their city. The upshot of this technology-speed-change vector for organizations is a universal call to innovate.

And the goal of innovation has changed. No longer satisfied with mere improvement, today's organizations seek to *reinvent* or *transform* themselves by adopting new digital ways of doing business—with new products, services, and delivery mechanisms—faster than their competitors, who increasingly appear from different industries. Until recently, grocery retailers did not consider Amazon a competitor. Neither did auto manufacturers. There is nothing like the threat of organizational irrelevancy to sharpen the focus on innovation! In fact, many organizational pundits suggest that innovation is the *only* form of sustainable competition.

This makes innovation a question of "how"—not "if"—for every organization. With technology on the critical path for almost all innovation, IT must reorient itself into a forward-thinking, business-oriented unit capable of marshaling resources to produce innovative business models that capture the attention of customers. In short, it is no longer "business as usual" for IT. The IT organization must develop the capabilities and mindset to drive innovation and thrive in the world of digitization.

The Innovation Challenge

In working with numerous CIOs and IT managers over the past decade in both the United States and Canada, we have studied many attempts at innovation. Some are "bet the farm" initiatives; others are experiments. But all involve vision, changes in mindset, and fortitude in adapting existing IT's conceptual and delivery mechanisms.

What we've learned is that innovation is a journey, not a destination. When innovation goes right, new questions and issues arise but organizations find a way to learn, adjust, and move forward. When innovation goes wrong, it leaves a bad feeling in the organization that can inhibit future change efforts. Successful innovation requires mental toughness, leadership, collaboration, and a willingness to stay the course—learning from what didn't work and strengthening and expanding on what did. As both the following examples show, innovation is not a superficial change for IT or for business, but a fundamental one.

When Innovation Goes Right. At an insurance company we know, the CIO successfully delivered a huge new enterprise initiative central to the company's transformation strategy by creating North America's first fully online direct insurance business, Concerto. For the first time, customers could obtain a fully binding quote for their car or home insurance online and purchase it without going through a customer service representative or broker to finalize the deal.

Although Concerto looked like an IT product, it was a business strategy and its success was due to full business participation. The project adopted a collaborative working model that facilitated problem-solving by means of agile and DevOps practices. Multidisciplinary teams worked continuously to develop, integrate, and implement new functionality using a hybrid model of cloud and on-premise technology that helped them respond to new demands and opportunities quickly. Accessing big data through application programming interfaces (APIs) helped create a sophisticated technological platform to deliver additional products.

Concerto brought the company's growth strategy to life, but it was just a start. Its leaders knew the next steps were to leverage the Concerto platform and experience for the rest of the business, especially the insurance brokers. The challenge was twofold:

- **Continuous improvement.** Financially, Concerto's premiums had grown, but its operations costs had grown even more. Since it wasn't just new business the company sought, but *profitable* business and business efficiency, attention needed to be focused on the balance sheet. Automating underwriting was an instrumental component in providing customers with a unique experience; however, the algorithms now needed adjustment to ensure costs were covered while remaining price competitive.

- **Applying what had been learned** to other company brands and processes. This involved further research, developing new ways of using data, new delivery mechanisms, new digital services, education, and cultural change.

When Innovation Goes Wrong. In contrast, at a bank we visited, the new CIO convinced senior executives they were missing opportunities to drive innovation with IT. Under the banner of "disrupt or be disrupted," funding was secured to establish an innovation center. The unveiling of the new facility, which was populated with giant screens, white boards, flexible spaces, floor-to-ceiling glass, and innovation "hives," created quite a buzz. Key business players were seconded to work closely with their IT counterparts to produce the first product—a slick standalone application that initiated loan renewals from a single screen with automated prompts, pre-filled fields, and knowledge bubbles for real-time learning.

The second product, meant to be another quick win, unfortunately stalled when operations and architecture both balked—operations because of lax security measures and architecture because of the nonstandard host language. Among the regular IT staff, the innovation center's "scrums and sprints" soon became code for "fast and loose." By the end of the first year, key business sponsors were being pulled away to fight other more important fires, and co-location had become more virtual than physical.

Partway through the second year of the innovation center's operation, senior executives began questioning the overall contribution to the business of what appeared to be a series of one-off projects that were difficult to reconcile with the enterprise's other strategic initiatives. The CIO, who was not one to acquiesce under pressure, suggested that the innovation budget needed to expand, not contract, in order to make significant investments in big data and analytics.

Just into its third year, the innovation center was repurposed to showcase the future of retail banking and the innovation budget had disappeared . . . as had the CIO.

Driving Innovation Successfully

In 2015, we decided to study innovation and IT by working with CIOs in Canadian and US organizations. We focused on topics that would foster improved capabilities—things like customer experience and customer centricity, strategic partnerships, digital leadership, strategic ecosystems, data analytics, intelligent organizations, and innovation funding models. Following up with these firms over this period revealed that their innovation success had varied considerably.

We would have expected all these organizations to have succeeded with their innovation agendas. Their IT functions were mature and their technology was advanced. Funding was not the issue. They had well-established planning procedures and experienced personnel. Yet some were more successful in driving innovation than others.

What we discovered was that their understanding of how to innovate also varied considerably. Some understood how to effectively move an opportunity from a concept to a practice and deal with the inevitable challenges involved. For others, there were problems—links that were missing, poorly functioning, or just misaligned—and these had inhibited the successful delivery of good ideas.

We learned that driving innovation successfully involves three stages—opportunity, discovery, and delivery (see Figure 1). And each stage requires new skills and capabilities. Some of these are complementary to more traditional business and IT skills; others are counterintuitive. We were then able to

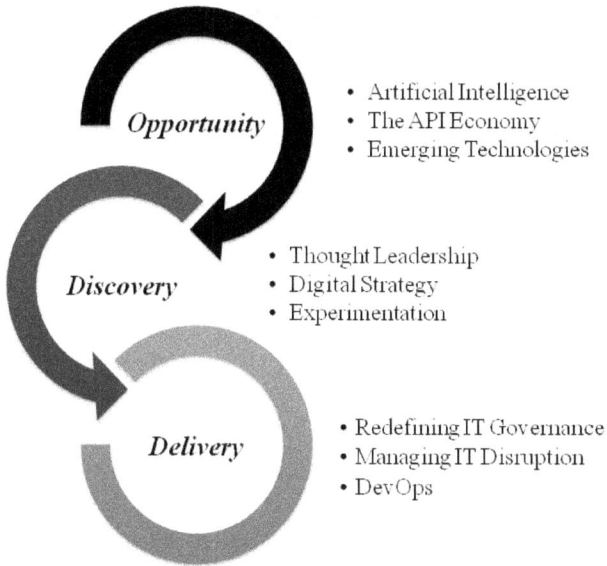

Figure 1. Innovation stages

bring together groups of senior IT managers to explore these innovation stages in detail to learn what is needed to drive them successfully and to identify where IT organizations face innovation challenges.

The result was a number of white papers describing the issues involved at each stage and suggesting workable solutions for some of the challenges IT organizations face in each. Taken together, they form a practical guide to successful innovation with which IT managers can identify and effectively address their own challenges.

Organization of This Book

This book is therefore based on the insights of many IT executives, and we believe it represents a collective wisdom that is invaluable to IT leaders grappling with how to enable innovation to proceed unimpeded. Chapters are grouped into three sets of drivers that represent the three distinct stages of innovation: opportunity, discovery, and delivery.

- **Section I: Opportunity Drivers.** These position an organization on the "big waves" of technological change while continuously monitoring and integrating other emerging technologies. Chapter 1, "Artificial Intelligence," explores the wave of artificial intelligence that has rapidly appeared on the horizon, for which all organizations need to prepare. It examines both the opportunities and challenges involved in adopting this transformational technology and how best to assess its value and manage its introduction and integration into an organization. Chapter 2, "The API Economy," looks at a less visible wave that will be no less transformative. All organizations must grapple with the possibilities that APIs now provide to tap into specialized services, such as cognitive computing and IoT networks, and business ecosystems. This chapter presents a framework for thinking about API usage in organizations and how leaders can get started with them. Chapter 3, "Emerging Technologies," presents a conceptual structure for identifying and assessing the new technologies that are relevant for an organization and then energizing organizations around imagining their possibilities.

- **Section II: Discovery Drivers.** Once an opportunity is identified, these drivers explore its potential value, enhance it iteratively, and situate it within an overarching digital strategy. Chapter 4, "Thought Leadership," explores the nature of thought leadership and the characteristics of an effective IT thought leader and discusses how IT functions can foster thought leadership successfully in their organizations. Chapter 5, "Digital Strategy," discusses the value of having a coherent digital strategy and the organization that is needed to support it. It describes how to develop one and looks at the broader implications of digitization for the organization and what will be needed to support a successful strategy. Chapter 6, "Experimentation," examines where, how, and whether experimentation can be used in IT and, if so, how it fits in with other new ways of working in IT. It then describes the experimentation life cycle and how this fits with more traditional IT delivery mechanisms.

- **Section III: Delivery Drivers.** Nontraditional approaches to technology use require new IT governance practices that ensure organizations can deliver innovation rapidly and continuously. At the same time, IT organizations must be restructured to cope with new ways of delivering technology. Chapter 7, "Redefining IT Governance," discusses how to evolve IT governance processes, which are often viewed as overly bureaucratic by the rest of the organization. It focuses on how best to balance the needs of the enterprise with responsiveness to business unit needs to ensure that IT can act faster and in closer alignment with business units while still protecting the organization's overarching interests. Chapter 8, "Managing IT Disruption," looks at the many different disruptive forces affecting IT today—both in the organization and in IT itself—and how these can be seen as opportunities instead of challenges. It concludes with a picture of what the IT of the not-so-distant future might look like as a result of today's disruptive forces. Chapter 9, "DevOps," explores what problems this new approach to delivery is attempting to solve, the value it can deliver, and suggests a plan for how to introduce it to the organization.

The Genesis of This Book

To learn how to drive innovation with information technology (IT) successfully, we held interactive sessions with senior IT executives and managers from a number of leading-edge organizations representing a wide variety of industries (e.g., retail, manufacturing, pharmaceutical, banking, telecommunications, insurance, food processing, government, and automotive).[1] Each session explored a specific innovation challenge they were experiencing. From these, we wrote reports incorporating good practices and guidelines augmented by relevant academic and practitioner materials for addressing the issues involved.

Over the years we have learned that these issues vary little across organizations. However, each tackles the same issue somewhat differently. It is this diversity that provides the richness of insight in this book, yielding a thorough understanding of each issue and strategies for how it can be managed successfully.

---

1 Thanks to ADT, Aviva, Bell Canada, BMO, Canadian Tire, CIBC, Dell-EMC, Education Quality and Accountability Office, eHealth Ontario, Elections Ontario, Empire Life Insurance, Fairmont Hotels, Grand & Toy, Greater Toronto Airport Authority, Liquor Control Board of Ontario, Loblaw, Metrolinx, Ontario Lottery and Gaming Commission, Ontario Teachers' Pension Plan, Ontario Universities Application Centre, Parmalat, P&G, Reliance Comfort, SCI Logistics, Scotiabank, and Sun Life.

Section I

Opportunity Drivers

Opportunity management entails positioning your organization on the "big waves" of technological change, while continuously monitoring and integrating other new technologies. Chapter 1, "Artificial Intelligence," explores how introducing artificial intelligence in organizations will require addressing "big picture" questions of change, as well as changes in data management, skills, digitization, testing, and audit. Chapter 2, "The API Economy," looks at the importance of APIs to innovation, identifies gaps in our understanding of how they are changing and shaping our economic and business models, and presents a framework for thinking strategically about APIs. Chapter 3, "Emerging Technologies," examines gaps in identifying and managing emerging technologies to improve how their potential value can be assessed.

Chapter 1

Artificial Intelligence

W e consider artificial intelligence (AI) to be a "big wave" phenomenon due to its potential universal application in today's organizations. The rapid evolution of AI in terms of machine learning, natural language, and neural networks combined with massive processing power introduce capabilities to potentially drive all aspects of commerce. Whether automating internal processes, picking stocks, predicting (perhaps manipulating) voting behavior, or routing traffic, the application of AI is limited only by our ability to imagine the future. As the analogy intends, just like surfers, organizations must position themselves on this wave of opportunity or risk missing an epic ride enjoyed by others.

Despite its long scientific history, AI has only recently become a hot topic for most organizations (Austin 2017; Austin et al. 2016; Andrews et al. 2016). Although AI has been *theoretically* possible for some years, it is due to the convergence of three conditions that it is now making it *practically* possible for its commercial use. First, powerful hardware is now accessible through cloud computing and super computers such as Watson (Austin 2017). Second, the huge amounts of data that drive AI are now available through apps, sensors, and other devices (Rometty 2016). Third, new software such as powerful algorithms, natural language, and machine learning have matured to a point where they can do certain jobs as well as or better than humans (Austin 2017; Rometty 2016).

Excitement about AI is growing in the business world (Andrews et al. 2016) and it is also beginning to creep into our daily lives. Whether it is autonomous driving, medical diagnoses, or stock picking, it is clear that AI is inexorably moving from the realm of science fiction to reality. But, as with any new technology, AI raises some important questions for how it is to be used in today's organizations. "We're really at the very early stages of this technology," said a manager. "But we know it's going to be incredibly transformational. We're just not sure exactly how."

It is widely agreed that AI will be disruptive in a number of ways. First, it is a platform shift—away from the traditional structured ways of doing business towards a more "natural" interaction that adapts to context and conditions (Austin et al. 2016). Second, it requires new ways of managing technology. It "learns" and is "trained"; it isn't programmed. It finds things we didn't know were there and arrives at outcomes we don't anticipate. It will need to be controlled, managed, and monitored in ways we aren't ready for at present (Andrews et al. 2016). Third, it replaces a considerable component of many knowledge workers' jobs and this will have repercussions for both organizations and society (Klotz 2016). And finally, when organizations cede some or all decision-making in an area to AI, they are going to have to address a number of legal and ethical issues related to privacy, accountability, governance, responsibility, and decision-making transparency, as well as a host of social policy considerations (Bernstein 2016).

In this chapter we explore some of these challenges and IT's role in selecting, designing, implementing, and managing AI. We start by reviewing our current understanding of AI, its strengths and weaknesses, and where it fits with other types of technology in use in organizations. Next we look at how organizations are using AI and where they see it will be valuable to them. Following this, we examine a number of dimensions of AI management that will be important during its introductory phase in organizations. Finally, we outline several recommendations for managers who seek to "get AI right."

What Is AI?

AI is the broad term that refers to many types of "smart" machines that enable organizations to tackle more complex problems than traditional, structured systems currently allow. It encompasses a number of subtypes, including:

- **Robotics**—where machines do physical work, which can range from manufacturing, to supply delivery in a building, to complex laboratory processing such as blood tests.

- **Machine learning**—where algorithms that improve automatically through experience are used for processing and decision-making (Quora 2017).

- **Neural networks**—inspired by biological neural networks and which aim to mimic brain connections. Machine-based neural networks have developed strong pattern identification skills and can also be a type of machine learning—discovering rules, developing new rules, and tolerating noise and variability in data.

- **Natural language processing**—which includes a variety of technologies that facilitate conversational interfaces between humans and a machine. The most well known is IBM's Watson that used these processes to win *Jeopardy* in 2010. Natural language processing is being introduced to the public through digital "assistants" like Siri and Cortana, and a variety of chatbots.

AI works differently from other forms of technology primarily because it doesn't function in the binary fashion ("if-then-else") used by modern computer programs and doesn't require fixed commands (Andrews et al. 2016; Austin et al. 2016). As a result, it can be used to *augment* human intelligence to make better decisions, analyze massive amounts of data, identify anomalies, and proactively predict events (Moore 2016; Rometty 2016; Hoffman 2016).

Data is the fuel of AI and one of AI's key values is helping us deal with the massive amounts of data currently being created—both structured and unstructured. "In many ways we live in an era of cognitive overload characterized by an exponential increase in the complexity of decision-making" writes one expert. "It's impossible to create protocols, algorithms, or software code to successfully anticipate all potential permutations, trajectories, and interactions" (Rometty 2016). The focus group agreed. "The driver in IT for AI is our data lakes and how we use them to provide value," said a manager. "As we improve with our big data and analytics initiatives, it is exposing more and more opportunities for how to use AI."

Although AI makes machines "smarter," the focus group stressed that today's AI applications are narrowly focused on a single function, such as image recognition, pattern identification, or a particular task. While they can be spectacularly good at these types of tasks, they still lack general executive functions and, because they are trained by humans, can incorporate human biases into their actions (Austin 2017). For example, Facebook's nudity recognition engine ran into problems when it banned pictures of breastfeeding moms and Michelangelo's statue of David.

As one expert points out,

Things that are so hard for people, like playing championship-level Go and poker, have turned out to be relatively easy for the machines.... Yet at the same time, the things that are easiest for a person—like making sense of what they see in front of them, speaking in their mother tongue, the machines really struggle with.... General intelligence is what people do ... we don't have a computer that can function with the capabilities of a six year old or even a three year old, and so we're very far from general intelligence. (Higginbotham 2015)

However, AI is evolving rapidly, the focus group noted. "It's all about the human-machine interface really," said one manager. "This line is moving. Twenty years ago, cheque recognition was cutting-edge AI. Today, we just take it for granted." The group also noted that the line between AI and other forms of technology is unclear. "At what point do we say it's AI?," one asked. He noted that self-driving cars use a variety of "AI-like" technology, such as image recognition to identify specific patterns (e.g., a stop sign), a rules engine for things that rarely change (e.g., what to do at a four-way stop sign), and machine learning that uses probabilities and judgment to determine if you should stop at the sign, and for how long, in order to avoid an accident. "Right now, if it's stuff we can't do, we call it AI," another concluded.

What Are Organizations Doing about AI?

Although of high interest to many organizations, AI has not yet made the leap into organizational practice, except in very experimental ways. Many members of the focus group were tinkering or testing AI applications, such as chatbots, but as one noted, "we haven't put them in charge of anything yet." "Right now, it's all fuzzy and experimental," said another. "We're testing right now so that we can learn and make sure it works," said a third.

Focus group members see AI as being part of a continuum that begins with big data, predictive analytics, and business intelligence and expands from there. "As we improve with these, AI will be part of a natural progression," said a manager. "But we believe existing data technologies can still improve our business before we need AI." Another noted, "We're still struggling with how to connect AI to our business." Nevertheless, exploration and experimentation are important to help both business and IT find the right role for AI and the right approach to investing in it. "One of the challenges is that AI is both a technology and a solution," said a manager. "And it flips between the two. We have to learn how to manage it effectively."

Another challenge for organizations is that "most valuable AI platforms are built on narrow, proprietary platforms, while most broad, general-purpose AI platforms lack ready-made valuable AI applications and require buyers in every enterprise to fund the redevelopment of new applications" (Austin et al. 2016). For this reason, most enterprises are not yet exploiting AI. "[They] want and need lower cost, lower risk, faster to deploy and easier to manage solutions that are built on a common technical infrastructure" (Austin et al. 2016). As a result, "The zeal of the possible is tempered by the practical," said a member. The focus group agreed that AI is still so expensive most organizations can't tackle it on their own. Therefore, they are relying on partners and cloud applications to acquire the solutions they are experimenting with.

Members also pointed out that, in these early days of AI, it is difficult to see the full shape of how it might be applied in the future. "Our implementations will be very different ten years from now," said a manager. "At this stage we're just using it to help us build faster processes," said another, "much like the car was initially envisioned to be a faster horse." As a result, many of the common applications of AI at present are in areas where massive amounts of data are required, such as with legal or medical information. In the focus group, two companies are using AI in this way to monitor their security logs for abnormalities. However, "if it recognizes a pattern that doesn't fit, it calls a grown-up (i.e., a human)," said a manager.

Another company is experimenting with using AI for basic underwriting using simple rule sets. "We're just taking baby steps right now," said a manager. "We need to get our data in shape, develop use cases, train the system up, and then reinforce it with more data." And one firm is exploring natural language voice response in its call center but again: "We need more data before it's successful." The focus group stressed that AI still works on the garbage-in, garbage-out rule so improving data and monitoring outcomes are paramount in their AI work. "We need to make sure it works because how would you know it's broken?," explained a manager.

Another organization is taking a different approach, working with IBM's Watson computer to see if it can identify data that might be relevant to its products and services and how it would apply. "We hope to use it to predict new products," the manager said. "We want to use it to differentiate insights from noise." Companies are also learning how *not* to use AI. "We have been looking at robo-investing," said a manager. "But we've learned so far that people don't really want robots making their decisions. We're therefore re-vectoring this technology to stress that it provides *advice* which someone can take or not."

Although companies are experimenting widely with AI for a variety of reasons, members believe that the primary driver is business cost reduction—involving people and processes. "In the long run, we can expect to see a decay of jobs over time as AI is used to replace not only factory workers but also knowledge workers," said a manager. And the cadence of change is speeding up, they said. As they look into the not-so-distant future, they see a world where low-level knowledge workers will be largely replaced by AI, where decisions of all types are made and/or supported by AI, and where data and data scientists will be kings.

Dimensions of AI Management

The focus group identified a number of areas related to AI management that their companies are beginning to explore. Members stressed that AI is not a project but represents a fundamental change in how work is done. As a result, the issues outlined below are merely some of the initial challenges that IT and business management should consider when preparing for AI. As these are addressed and as AI evolves, new dimensions of AI will undoubtedly arise.

- *Digital transformation.* Most of the focus group companies are well down this path, which involves embracing new and different technologies in ways that challenge operational and value assumptions and integrate them with existing technologies to deliver new products, services, business models, revenue streams and/or customer/stakeholder experiences. What many may not have grasped yet is that "digital is not the destination. Rather it is laying the foundations for a much more profound transformation to come. Within five years . . . all major business decisions will be enhanced by cognitive technologies" (Rometty 2016).

- **Data management.** There is no question that AI requires more and better data to make it effective.

  Data is the lifeblood of AI. To train computers to learn ... you have to feed them tens of thousands of examples of something. The computers try to understand what elements of those examples define what makes a cat a cat in an image or what gives meaning to a certain word. The algorithm then gives a statistical weight to each guess that helps the computer "learn" what the right answer is. The computer scientist helps train the algorithm by giving feedback and more examples along the way. (Higginbotham 2015)

  "AI is forcing better data management," said a manager. One writer suggests: "Data [is] the world's great new resource. What steam power, electricity, and fossil fuels did for earlier eras, data promises to do for the 21st century—if we can mine, refine and apply it" (Rometty 2016). However to do this, companies need to build strong data functions and address the perennial problems of data ownership, privacy, security, and data classification, as well as the newer areas of big data and its management.

- **Business value.** As with other new types of technology, organizations want to understand how to use AI to deliver value. This value could come from replacing or augmenting human labor or from new products and services that have yet to be conceived. At present, organizations are mostly exploring how other firms are using AI in applications or in specific industries and identifying areas in their own firms that could benefit from AI use (Andrews et al. 2017). Supporting human decision-making is a primary area of interest (Andrews et al. 2016). The focus group added that working with big data and business rules exposes potential opportunities for AI. "We should attack these opportunities as they arise but also allow for synergies to develop and serendipitous sources of value to surface," said a manager. Others pointed out that to truly deliver business value, business practices will likely have to evolve as well. Although it will be IT that will make AI happen, the members agreed, AI implementation must be an enterprise-wide initiative with CEO sponsorship and funding. "AI is going to happen quickly and everyone must be together on this," said a member.

- **Skills development.** There is general agreement that we really don't know much about the specific skills that will be required to work with AI (Austin et al. 2016; Bernstein 2016). What we do know is that they are scarce. The focus group believes that data, analytics, data mapping, quality control, and quantitative measurement will be key emergent skills. "We will also need algorithm and modeling skills," added a manager. Methods for working with AI do not yet exist and this inhibits organizations' understanding of what skills to look for (Austin et al. 2016). At present, the best advice is to seek broad problem-solving skills and the ability to work on fluid teams (Bernstein 2016). In the shorter term, Gartner predicts that by 2019 more than 10 percent of IT hires in customer service will be writing bot scripts (Andrews et al. 2016). The focus group was optimistic about their organizations' abilities to acquire AI skills. "We have some staff already who have AI training and who are eager to use it," said a member. "And if we ask, our people will want to learn new skills." Members stressed the importance of having both business and data expertise in the future, as well as technical skills. "Our skill sets will change and entry level skills will be replaced with automation, but we will train at a more senior level so humans will still have 'skin in the game,'" one concluded.

- ***IT's role.*** IT's role in the organization will change because it is likely that responsibility for AI will be split between different parts of the enterprise (Andrews et al. 2016). The focus group agreed that IT's role will be primarily that of creating the right conditions for enabling AI and integrating it with existing systems. That said, IT will play a part in many dimensions of AI work including: helping to identify opportunities, clarifying the purpose of an AI implementation, cleansing and identifying data, managing and maintaining the AI environment, helping to select and implement appropriate algorithms and maintaining them, ensuring quality outcomes, building bridges between applications, and coordinating with privacy, legal, and security groups. In addition, it is clear that current IT practices will need to evolve with the advent of AI, particularly enterprise architecture, vendor selection, software development, and business intelligence (Andrews et al. 2017). AI implementation is best suited to iterative development, noted the focus group, so it is important to develop a competence in this area. Finally, it will be essential to create a governance structure that will enable decisions about AI to be made effectively and to ensure accountability for these decisions. The focus group anticipates that, at minimum, an AI implementation will need several sets of approvals from business, IT, legal, HR, and advisory groups.

- **Testing and audit.** Testing is especially critical when working with AI because AI can detect unique and unanticipated patterns in data (Austin 2017). Therefore, traditional methods of testing all code paths aren't sufficient. AI outcomes must be monitored over much iteration with multiple data sets. Sometimes, with the best of intentions, AI simply yields a wrong result. It is therefore the responsibility of testing to ensure that no wayward results occur. Experts caution against the assumption that, once deployed, smart machines will need no further attention (Austin 2017). This is a fallacy that needs to be corrected. AI, for all its strengths, will need to be retrained and retested as new data are collected. Failing to consider "the challenges of continuously maintaining and monitoring an implementation will lead to failure in many enterprises" (Austin 2017). The focus group also noted that many industries require a clear audit trail that will justify decisions made and that can be used to root out discrimination. "Whether or not we participate in decision-making, we are still responsible for those decisions," said a member. "Therefore, where a decision is important, it's essential to have an audit trail built into all AI algorithms."

- *Cognitive ergonomics.* The focus group noted that any AI implementation must fit into an organization's social fabric as well as address specific business opportunities. One of the biggest emerging issues in AI is therefore how this technology will interact with humans. In a world where machines and humans will collaborate on problem solving, decision-making, and customer services, organizations need to find the right way to blend humans and machines successfully (Andrews et al. 2016). Cognitive ergonomics is a new field that considers the why and how of AI implementation, taking human and social factors and design principles into account.

Driving Opportunity with AI

One of the most challenging aspects of AI is that it is leading managers and society at large to consider some of the broader impacts of technology adoption. "There are real social impacts to this technology," said a manager, "and we need to adopt it in a way that is mindful of them." The focus group identified five big issues that need to be tackled when adopting AI:

1. *Ask the bigger questions.* "AI raises many questions that still need to be resolved," said a manager. Within organizations, the focus group easily listed a number of issues that are being discussed by IT leaders, including:

   - Should a bot self-identify as a robot?

   - Should AI be transparent about how it makes decisions? Is informed consent needed to use AI?

   - Who accepts decision responsibility and accountability? The organization? The algorithm supplier? The data scientist?

   - Should waivers be required in some cases? Are they ethical?

   - What are some of the "back door" implications of AI (e.g., smart TVs that leak data)?

   "We don't understand what norms for using AI will be acceptable," said a manager, "and these will likely vary around the world. They are likely to evolve more slowly than the technology itself."

There are even bigger questions that must also be addressed. AI and robotics are beginning to affect labor markets (Bernstein 2016), and most predictions point to increasing levels of job losses over the next ten years (Hoffman 2016). One writer notes, "The disconnect with past work models is happening a lot faster than in the past. . . . We'll soon see enormous waves of workers put out of work and ill prepared to take on very different jobs" (Bernstein 2016). The focus group was very aware of the potential societal dangers involved in such massive economic displacement and the fact that our institutions are ill-equipped to deal with these changes. "Our welfare, unemployment, retirement systems and our universities all need to adapt," said a manager. Another added, "We need to work to maximize 'friendly' AI to extend human intelligence and open new fields of employment." "There are real social impacts to AI and we all need to work together to identify the questions that need to be asked, establish norms for its use, and reform our social and educational institutions," a third manager concluded.

2. ***Beware of anthropomorphism.*** "Anthropomorphism is the attribution of human traits, emotions, and intentions to non-human entities and is considered to be an innate tendency of human psychology" (Wikipedia 2017). As computers become more and more human-like in their ability to interact conversationally, it is

natural to ascribe human characteristics to them. AI developers try to leverage anthropomorphism to make computers easier to use. While not necessarily unwise, experts warn that the inappropriate use of anthropomorphic metaphors creates false beliefs about the behavior of computers such as overestimating their "flexibility" (Wikipedia 2017). For example, a customer service call center with "chatbots" could lead to disaster if it is unable to address complex human needs. This is a real danger, said the focus group, when companies are under constant pressure to reduce costs.

3. ***Work to develop trust.*** It is critical that people and organizations be able to trust what technology is able to do (Austin et al. 2016). At these initial stages of AI, this trust must be constantly tested. Organizations can expect vendors to oversell their capabilities as well, leaving people skeptical of what AI can really do (Andrews et al. 2017). Furthermore, trust will vary by context. One manager noted: "The level of trust required depends on the types of decisions AI is making—less is needed when determining the best route to work—much more when there are safety implications." In addition, trust can be misplaced or abused and this should never be forgotten. "We had total confidence in our automated airplane tracking system until Malaysia Airlines Flight 370 completely disappeared. Such occurrences reveal inappropriate assumptions that are

temporarily threatening and require a complete reassessment of how we are using technology," stated a manager.

4. ***Build multiple work models.*** As the world changes in response to new work practices resulting from adoption of AI, organizations will have to integrate their legacy systems and practices into this fast-paced, rapidly changing environment (Bernstein 2016). No one knows what models will be effective in this new world, so the best advice is to experiment with multiple ones (e.g., crowdsourcing, distant manufacturing or transaction processing, and contract work). Experience with different work models will help develop flexibility and agility and start to modify organizational cultures (Bernstein 2016). Having an adaptive culture will give organizations much more than a one-time advantage. It could be key to their very survival. One manager quoted Charles Darwin: "It is not the strongest of the species that survives, nor the most intelligent, but the one most responsive to change."

5. ***Consider open AI.*** Open AI is a non-profit artificial intelligence research company, supported by Tesla's Elon Musk, which aims to carefully promote and develop friendly AI in order to benefit, rather than harm, humanity as a whole. It is also an open-source

project aimed at creating specifications for AI and associated programs and tools. Its short-term goals are to build tools and algorithms that will be shared publicly and longer term, to develop better hardware that can perform more like a human. An open-source model is a cheaper way to address AI problems, and if it works it could help advance AI for everyone (Wikipedia 2017; Higginbotham 2015).

Conclusion

AI has now moved beyond the realm of science fiction and is just about ready for prime time. It is appropriate to ask ourselves important questions about how it could be used wisely or unwisely in organizations, and what needs to be done to mitigate the larger social impacts it is likely to cause. Preparing for AI is a daunting task. Thoughtful business and IT leaders must not only consider its potential value, but also its broader costs. Anticipating that economic pressure will eventually force AI adoption, organizations should seize the opportunity now to educate themselves about the different ways they can deploy it, and develop principles for its use that will take the larger social context into account. In addition, it is incumbent upon all organizations to work collaboratively with governments and researchers to ensure that the negative impacts of AI are addressed and remediated.

References

Andrews, W., F. Karamouzis, K. Brant, M. Revang, M. Reynolds, J. Hare, and D. Berman. "Predicts 2017: Artificial Intelligence." Gartner Research Report, ID: G00317025, November 23, 2016.

Andrews, W., D. Berman, A. Linden, and T. Austin. "Artificial Intelligence Primer for 2017." Gartner Research Report, ID: G00318582, February 3, 2017.

Austin, T. "Smart Machines see Major Breakthroughs after Decades of Failure." Gartner Research Report, ID: G00291251, January 4, 2017.

Austin, T., M. Hung, and M. Revang. "Conversational AI to Shake up your Technical and Business Worlds." Gartner Research Report, ID: G00315689, September 30, 2016.

Bernstein, A. "Globalization, Robots, and the Future of Work: An Interview with Jaffrey Joerres." *Harvard Business Review*, October 2016, 74–79.

Higginbotham, S. "Here's Why Elon Musk and Everyone Else is Betting on AI." *Fortune*, December 16, 2015.

Hoffman, R. "Using Artificial Intelligence to Set Information Free." *MIT Sloan Management Review* 58, no. 1 (Fall 2016).

Klotz, F. "Are You Ready for Robot Colleagues?" *MIT Sloan Management Review Digital*, July 6, 2016. http://sloanreview.mit.edu/article/are-you-ready-for-robot-colleagues/.

Moore, A. "Predicting a Future Where the Future is Routinely Predicted." *MIT Sloan Management Review* 58, no. 1 (Fall 2016).

Quora 2017. "Machine Learning." https://www.quora.com/topic/Machine-Learning (accessed March 29, 2017).

Rometty, G. "Digital Today, Cognitive Tomorrow." *MIT Sloan Management Review* 58, no. 1 (Fall 2016).

Smith, H. A., and J. D. McKeen. "Developing a Digital Strategy." Smith School of Business, Queens University, IT Management Forum, 2015. https://smith.queensu.ca/it-forum/index.php.

Wikipedia, 2017. "Anthropomorphism." https://en.wikipedia.org/wiki/Anthropomorphism.

Chapter 2

The API Economy

With the advent of mobile technologies, we have all realized how connected we are. But we don't often realize how this connectivity takes place or where it is leading us as individuals, organizations, or economies. With the addition of the Internet of Things (IoT), there will soon be literally billions of devices in our homes, businesses, and elsewhere all wanting to interact with each other, generating zettabytes of data (Susain 2016).

This connectivity is increasingly being driven by application programming interfaces (APIs), which we consider the second "big wave" phenomenon because of its universal impact on organizations. The combination of IoT and APIs will leverage innovation in countless ways unfathomable even five years ago. The basic idea of an API is that the owner of one application creates a set of access methods that can be called by another application. The API is documented and, if used correctly, it creates a level of abstraction between the two applications. Through this means changes can be made to either application without affecting the way they interact.

This is not a new concept in IT. Programmers have used something similar for several decades to standardize communication between a system's modules, thereby isolating chunks of functionality and reducing errors (Clark 2016). However, APIs have gained new attention in recent years because technology start-ups (unburdened by legacy systems) have used them to achieve faster time-to-market and create new functionality by accessing data and chunks of functionality (e.g., Geolocation) *externally* as well as internally. As a result, larger organizations are now increasingly relying on external APIs to help them do the same (O'Neill 2016).

But APIs are more than basic tools for application developers and data scientists. By enabling the rapid reuse and recombination of disparate data and functionality, they also facilitate the development of completely new products and services, and they allow companies to participate in ecosystems of organizations without the traditional need for extensive negotiation and customization of information systems. In addition, they are now enabling companies of all sizes to tap into specialized services such as cognitive computing and IoT networks that are too difficult, expensive, or time-consuming to reproduce in-house (Narain et al. 2016).

These new capabilities enabled by APIs mean that APIs are now not just a technical element of IT, but they have become a strategic business priority that will change the nature of organizations and business models as well (Collins and Sisk 2015; Narain et al. 2016). And as the commercial exchange of business functions and capabilities using APIs expands, it is predicted that an API economy will also evolve to utilize them more effectively. But today, understanding how and where to use APIs remains a challenge for both business and IT leaders, and despite the hype there still is much to learn about the effective utilization of APIs to deliver business value.

In this chapter we explore how organizations are approaching the opportunities and challenges of using APIs. We begin by better defining what an API is, where it fits in with other IT work, and how APIs are expected to deliver value. Then we examine how the use of APIs is anticipated to lead to economic transformation. Following this, we present a framework for thinking about API usage in organizations and offer some practical advice for how leaders can get started with APIs.

What Is an API?

APIs are multipurpose tools that provide simple external interfaces for a variety of purposes (Clark 2016). The focus group explained that an API is an arm's length, shareable interface that acts as a common communication channel providing access to data and capabilities. "In the past, business logic and presentation were combined in a single system," said a manager. "Now, we keep channels separate from logic." An API is therefore a standardized software component acting independently of its host application and enabling bridges to be built between applications (whether inside or outside the organization) to bring together disparate functionality to create new forms of value (Malinverno 2016).

The term "API" is often used interchangeably with "SOA" and "microservices." Although some have attempted to distinguish between these, the fact is that "it is impossible to gain agreement on how they relate to each other." The tools are merging and combining in our thinking (Clark 2016). One focus group manager noted that "microservices are a way of building fine-grain composable functionality and APIs are the interface to it. Microservices talk through APIs." Others simply state, "APIs expose assets like data, algorithms, and transactions [and] make it easier to integrate and connect people, places, systems, data, things, algorithms, to create new products/services and business models" (Collins and Sisk 2015).

It is more helpful to distinguish APIs based on the functionality they enable:

- **System APIs** provide a means of communicating and interacting with legacy systems. For example, a company may need to call a balance transaction from legacy system to use in its mobile app. System APIs remove the need to know specific protocols and/or business logic unique to the underlying legacy systems.

- **Business APIs** are composable services that include logic, such as "customer lookup," or "provide account summary," or "present a consolidated view." They offer hybrids or combinations of services and could be internally or externally focused.

- **Experience APIs** shape services for a user interface (UI), such as for a mobile app. "These emphasize how we engage with customers," said a manager. "They're not about core business logic but about shaping data to a UI and speed of response.

- **Algorithm APIs** provide access to a particular piece of computational logic.

Until recently, APIs have been built on a one-off basis by individual project teams (Collins and Sisk 2015), and it has often been assumed that APIs are easy to consume by others (O'Neill 2016). The reality, however, is that APIs add new layers of complexity to IT services, and companies are now looking for more formal ways to manage them. Organizations are also concerned about the risks involved of becoming more connected online in myriad ways (Longbottom 2015). Thus there is a need for API management tools (Columbus 2017; Golluscio et al. 2017).

Three types of tools simplify the consumption and use of APIs:

1. ***API portal.*** A portal gives developers the ability to discover APIs and experiment with them (Clark 2016). If APIs are exposed to external developers, they also provide a means for developers to register to use APIs and pay them if they develop a useful product/service using organizational data/functionality. In addition a portal provides protocols and policies for allowable interactions.

2. ***API mediation.*** This manages the authorization and security between the calling and responding functions of APIs, as well as managing audit and information transformations (Longbottom 2015; Malinverno et al. 2017). It supports all API interactions (i.e., between developer and external APIs and between IT systems and devices). API mediation also ensures security (including authentication, authorization, and protection), traffic management, and orchestration so that APIs can be customized for different constituencies and usage monitoring (Golluscio et al. 2017).

3. ***API portfolio.*** Since APIs are not transient, they must be managed over their life cycle (i.e., establishing a clear definition of value, a defined audience, and measurement of effectiveness) (Collins and Sisk 2015).

Because they enable an organization's core assets to be reused, shared, and monetized and thus extend the reach of existing services and possibly create new revenue streams, APIs are now considered a business model driver worthy of boardroom consideration (Collins and Sisk 2016). And by creating a platform for digital commerce, APIs are the foundation for every digital strategy (Malinverno et al. 2017). However, APIs and API management are still in their earliest stages and companies have much to learn about how to use them effectively and how to fit them into their broader business strategies (Columbus 2017).

The Value of APIs

An important factor in being successful with APIs is an understanding of their potential value and how and where they should be used. Although there is a great deal of hype in business about using APIs to sell corporate data and generate more revenue, the reality is that "few companies have mastered these capabilities and they take years to develop. Many companies *want* to do this but the problem is that it's very difficult to do" (Wixom 2016).

Much of the confusion about APIs stems from complexity about the different ways they can be used to deliver value. There are four primary ways that value can be derived from APIs:

- ***APIs for improvement.*** Due to their enhanced ability to integrate applications, APIs can streamline development processes and dramatically reduce an organization's time-to-market with new products (Columbus 2017). These advantages help organizations improve customer experience, comply faster with new regulations, and rapidly expose their products and services to the broadest possible audience (Collins and Sisk 2015).

- ***APIs for leveraged products and services.*** A second way of using APIs is to differentiate a product or service by enhancing it with data, new forms of presentation, or new functionality (Wixom 2016). This is done by providing access to some of an organization's APIs to carefully vetted external developers. These help companies offer an improved customer experience, add new digital products, and open new business channels to the market (Narain et al. 2016; Malinvero et al. 2017). They also enable the orchestration of a number of different APIs to facilitate new business processes while providing real-time integration. For example, software vendors now provide organizations with a number of common insurance functions through APIs, such as VIN lookup, address verification, and real-time verification of insurance. A key component of delivering value in this way is having a clear understanding of customer needs because these must be reflected in the APIs and the apps that are subsequently created (Columbus 2017).

- *APIs for interorganizational innovation.* Organizations may also seek "frictionless" transactions with trusted partners to leverage each other's data and services in order to reduce costs and time (Narain et al. 2016). To do this, a company offers a subset of APIs to its business partners. The more business-sensitive the API, the more tightly these partners are vetted and managed (Malinverno 2016). The value in sharing APIs may come from selling data, attracting new customers, or retaining existing ones through competitive differentiation.

- *APIs for continuous platform innovation.* The most visionary form of API value derives from opening up company data and services to a much broader and more open ecosystem of developers and organizations to enable them to create radically new products, services, and business models. Predictions are that APIs will be "the new conduits through which future innovation can and will be realized globally and drive the next level of differentiation" (Narain et al. 2016). In this scenario, a company creates a platform of APIs that supports the creation of an external ecosystem with connections to new marketplaces and communities. These APIs open new business channels, bring in more clients, maximize client retention, and enable the development of apps a company either doesn't have the time, ideas, or the resources to develop (Malinverno et al. 2017). Here APIs may be saleable products that generate revenue every time they are "called" (Clark 2015). An example of this type of API use is Salesforce, which uses APIs to jump-start new solutions and offerings from other developers.

Each of these approaches to API value requires different capabilities and organizational commitments to strategy, design, and execution (Wixom 2016). To better understand these, leaders should ask the following questions about any API initiative:

- How is value created with this API?

- How is value measured?

- Who owns the value generation?

- Whose problem is solved?

- What are the key risks?

The API Economy

The API economy is a catchphrase suggesting that APIs are a rapidly expanding economic force (Longbottom 2015). It is characterized by a "marketplace driven by data that uses APIs to reach customers" (Narain et al. 2016). In this economy, APIs act as the digital conduit linking services, applications, and systems. They enable organizations to share data and applications using easily accessible standards and platforms. These, in turn, allow businesses to make the most of their data to create compelling customer experiences and open new revenue channels. In short, the API economy is the commercial exchange of business functions and capabilities using APIs. It has captured the attention not only of software developers, but also of strategists and business leaders seeking to move to the next level of marketplace differentiation (Narain et al. 2016).

What this means is that organizations won't act as lone entities anymore. In the API economy, companies will work together to create more value than either of them could independently (Anuff 2016). Moving forward, APIs will redefine the nature of partnerships, allowing companies to collaborate without the traditional need for extensive negotiation and customization of systems (Narain et al. 2016). APIs are also the enablers that turn individual businesses into platforms (Pettey 2016). As one researcher states, "It's not enough for a business to serve its customers and make money, it has to be a platform and you can't be a platform without APIs" (Anuff 2016). And it's not an option. Focus group participants noted that APIs are being forced on them in a variety of ways. "Our regulators are now requiring us to provide APIs," said one member. Another stated, "In banking, if you want to participate in foreign exchange, you must use them." "APIs have already externally changed the game we're in," said a third.

In our present economy, products and services from a supplier are pushed to customers either directly or through intermediaries (Isckia and Lescop 2015). In the future, APIs will evolve from enabling simple connectedness to supporting remote interactions across a network, to platforms where APIs facilitate and accelerate new service development, to ultimately becoming the actual product or service a company delivers (Collins and Sisk 2015). As such they will become the fuel that keeps companies competitive and drives a significant economic shift. In an API economy, the products are APIs and the market is global (Malinverno 2016).

The API economy can thus be viewed as a set of business models and channels that provide secure access to functionality and data (Malinverno 2016). These create a platform that attracts partners who will develop and market its products (Isckia and Lescop 2015). Together, all entities affiliated with a platform act as a business ecosystem that builds on the strengths of others, takes advantage of shared affiliations, and draws in new participants.

By connecting people, businesses, and things into digital platforms through APIs, the API economy will be driven by a different economic logic. Platforms will serve as mediating entities that create value by facilitating interactions between agents that operate on different sides of a digital market (Hoelsch and Ballon 2015). And the pace of change will constantly accelerate through recombinations of resources and knowledge. A platform's success will depend not only on the platform's owner but also on its members' ability to innovate.

When a platform has attained a critical mass of participants, entry barriers will be high. Competition will be about who has the best platform strategy and the best ecosystem to back it up (Isckia and Lescop 2015). Differentiation will come from the APIs available and how motivated developers are to create applications using them (Malinverno et al. 2016).

A Framework for Thinking about APIs

Focus group companies were in the earliest stages of thinking about API strategy. "At present, we are driven by demand," said a manager. "If there's a market we want to participate in, we must use APIs." Others noted that much of their API use is motivated by internal productivity. "We've justified API use by reusability and simplicity," said a member. "Our business leaders believe that APIs mean reuse which means go fast." Another added, "Large companies also have internal customers and they have the same types of problems as our external customers. Sharing data internally can deliver real value. We're gaining tons internally."

They admitted they were struggling to articulate a comprehensive API strategy. "We need to figure out what data we're going to expose both internally and externally and how we are going to manage both levels," said a manager. "It's a struggle about where to focus because of the different layers involved," added another. In addition to the strategic complexity involved, organizational leaders are debating the risks of using APIs in external marketplaces. "Our executives are scared of external exposure," stated a member.

There is broad recognition that companies must begin to think about external APIs, and the focus group had numerous ideas on how to do this. Some want to create their own marketplaces that attract others to their platform. This model is particularly appealing to industries with a few large players and lots of peripheral development, as in banking, where there are lots of small Fintech companies experimenting and innovating with particular banking functions. Others are looking at externally sourced data to determine how they can incorporate it into their own apps and systems to make them richer and more contextual. Some are exploring how they might best monetize their own data for external consumption. Companies are also exploring partnering as an alternative to fully open markets. "We're using APIs primarily as a means of integrating with software-as-a-service," said a manager. Still others are focusing on making themselves easy to work with through APIs. "It's all about selling more stuff. Competition is easier if you are easy to integrate with," said a manager. One manager concluded, "We have no major external drivers. APIs feel more like a survival strategy to us."

Most large companies are approaching APIs cautiously while recognizing that APIs and their management must be tackled. If the four types of API value are configured according to whether they enable broad use of an organization's APIs or a more limited set, and whether they use a closed (or tightly managed) API market or an open one, an *evolutionary* framework emerges. This traces API use from limited, closed, internal approaches to more controlled external uses to more wide-open strategic uses (see Figure 2.1). Companies can then consider how best to incorporate each component of this framework into their overall API strategy.

- *Closed API market; limited API use.* This stage uses APIs to help organizations improve internally. Improving APIs internally requires thinking about improving data quality and standards and decomposing legacy systems into pieces of functionality for ease of access. Thus this work helps build data and service management skills and acts as a foundational piece for other API work (Wixom

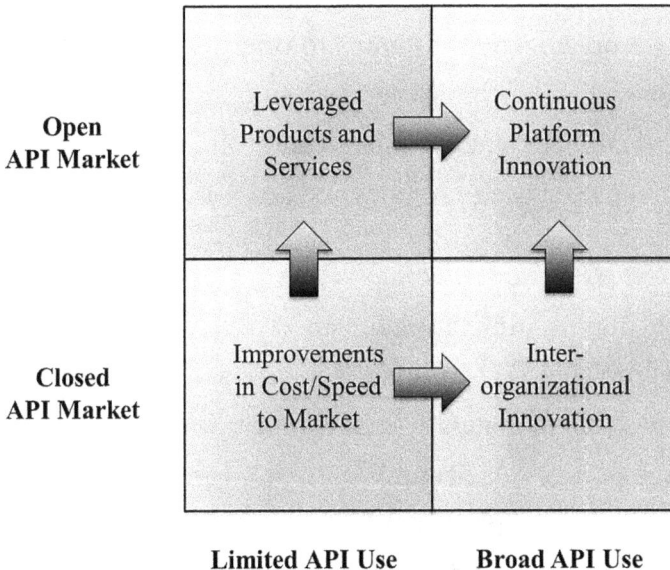

Figure 2.1 An API usage framework

2016). Due to their substantial legacy environments, most large companies tend to start at this point.

- *Open API market; limited API use.* As companies gain experience and confidence with APIs, they could decide to become a modular producer of plug-and-play products or services. Providing access to a limited set of APIs enables external developers to add value to an organization's products and services. This stage helps a company enhance its customer experiences in a variety of ways and creates new channels, enhanced products, and broader markets (Weill and Woerner 2015).

- *Closed API market; broad API use.* This stage uses APIs to enable a limited number of partners to more broadly share data and functionality (Anuff 2016). Here organizations attempt to offer real-time integration and build a limited ecosystem of trusted companies with complementary skills to add value and help them differentiate their products and services.

- *Open API market; broad API use.* This stage supports continuous innovation in an ecosystem (Isckia and Lescop 2015). It provides a branded platform and uses APIs to generate sales through both partners and third parties (Weill and Woerner 2015). It supports the continuous implementation of new configurations of products, market approaches, processes, technologies, competencies, and management systems. It also represents a new way of solving problems and discovering new business opportunities, leaving them to the external marketplace to solve. It also delivers new revenue streams through monetizing API use and new sources of value by delivering new groups of customers. Focus group members noted that this approach is more frequently used by "tech companies" such as Google or Netflix. "These are not representative of our business models," said a member.

This stage is the most challenging to implement and carries the most risk. However, it is also the most rewarding (Narain et al. 2015). It is very important to get the value proposition right and to remember that the ecosystem must provide sustainable incentives and rewards to API providers, developers, and users (Anuff 2016) (see box).

Regardless of the starting point, there are a number of common questions that companies will need to address when beginning to work with APIs, although these may have different answers depending on the approach(es) used. These include:

- Which of our platforms need APIs? How can we improve our understanding of the different approaches involved?

- How should we establish *options* for our future?

Doing Business with APIs

Of the top 50 most downloaded apps in Apple's app store, only a handful are fully functional with Siri.... What happened? Many developers chose not to sign onto the [APIs] ... to integrate the assistant ... because Apple only let Siri be used in six categories ... and that knocked just about everyone out.... It was just too restrictive....

Slow uptake [of APIs] from app developers risks further denting Siri's credibility, already bruised by the growth of Alexa [and] Google Assistant.... Siri is struggling as other assistants get smarter.... Apple's rivals have gotten developers on board where Apple hasn't. Alexa ... just passed 15,000 available "skills" ... [while] ... Apple lists fewer than 100 ways to use Siri.

(Adapted from Graham, 2017)

- If we are building external APIs, which internal APIs might they depend on?

- How can we motivate and manage both internal and external developers using our APIs?

- How can we attract and build partnerships to enhance our API strategy?

- How can we best integrate complementary products?

- What monetization strategy should we use: free; pay per use; tiered access; revenue sharing; subscription; or premium access?

- How will our architecture need to change with our API strategy?

- How should we assess the quality and value of an API?

- How will we manage an API's life cycle over time?

Driving Opportunity with APIs

It's clear that APIs will require a lot of change not only in business strategy, but also with the more practical aspects of how APIs are conceived, developed, managed, and governed. Managers therefore need some practical ways to think about how to get started using APIs. The focus group had the following advice for other business and IT leaders:

1. *Identify potential sources of API value.* First and foremost, organizations should seek to solve their customers' problems. This step transitions an organization from 'know your customer' to 'know how your customer needs to change'; from making products easy to use to understanding customers' behavioral needs; from monitoring usage to monitoring value created; and from identifying unrecognized needs to converting

unrecognized needs to 'must haves' (Wixom 2016). Having a strong focus on the customer and creating customer-centric APIs are essential to the success of any API initiative (Columbus 2017). This means taking an outside-in perspective rather than focusing on internal complexities or organizational siloes (Collins and Sisk 2015).

Sharing with APIs can meet strong institutional resistance. Ultimately, the focus group stated, it is the business that must decide whether this sharing will help or hurt, but leaders should ensure that their organizations take a longer-term view of their customers and understand what is and is not differentiating about their business.

2. *Develop API governance.* As a new element of an organization's business and technology strategy, APIs have mostly been adopted on an as-needed basis by whomever wishes. As a result, the focus group noted that there has been a proliferation of APIs without much coordination and oversight. API governance serves many functions. First, it ensures that new APIs are aligned with business strategy (Maliverno et al. 2017). Second, it establishes management, coordination, and control over API use by putting registration, cataloging, and monitoring practices in place. Without these, there is a risk of sharing private or mission-critical data unwittingly (O'Neill 2016). Third, it makes common

investments in the organization's platform to balance the tensions involved—between control and creativity, standardization and variety, and individual and collective needs. In short, it acts as the regulator of the organization's platform (Isckia and Lescop 2015).

A key debate in many focus group organizations is about API ownership. Members agreed that the business owns APIs but felt that recognition of the overarching issues and interdependencies involved was generally missing in business. They advised being pragmatic about how much control to exert, at least at these early stages. "The best thing to do at first is to monitor how APIs are developed and used and to spend governance dollars on remediation," said one member. "Plan big and start small." Initially, a governance group should seek to identify the APIs currently in use, and understand their usage and what agreements are in place (O'Neill et al. 2017). From here, it can work to reduce redundancies, rationalize providers, optimize traffic, and reduce costs (O'Neill 2016). By treating APIs as corporate assets that need management throughout their life cycle, governance emphasizes their importance to the organization (Maliverno 2016).

3. *Change how you think.* "The most important change we in IT need to make is in our minds," said a manager. "Traditional systems development is like the soviet

economic model—we will decide what you use; APIs on the other hand are more capitalistic because there's a marketplace." This mindset shift must extend to both strategists and developers. With APIs and microservices, software becomes the creative composition of pieces of functionality and data that can be pulled together quickly rather than a monolithic application (Narain et al. 2016).

4. *Adopt new tools and capabilities.* Although not every project will require the use of APIs, those aimed at delivering value or new capabilities, or at innovation and exploration, will need to use new methods of development such as agile techniques and DevOps, as well as new API-oriented standards and methods (Gilpin and Marshall 2017). Initially at least, IT will need to become bimodal—operating in both new and traditional ways (Maliverno et al. 2017).

There are two approaches to building API capabilities: building them on a product-by-product basis, or building an internal API practice that creates APIs strategically (Gilpin and Marshall 2017). At present the companies in the group are focused on learning how to build and use APIs on a one-off basis, but members recognized that more effort should be taken in the future to design and coordinate APIs, ensuring that they expose the right data and functionality. "We have started with a vocabulary about APIs," said a manager, "and established standards and component names." This includes a common understanding of the data involved (e.g., a credit card number), and published definitions of what an API needs. Focus group managers suggested that organizations also need standards and protocols, methods about how to build a good interface, documentation guidelines, and clarity around service level agreements in order to develop and use APIs effectively. And if an organization is using external developers, it must have a means of registering and controlling their access to company APIs. "We need guidelines and documentation," said a manager, "or we're going to have a literal forest of APIs. Ideally, this should be done at the enterprise level, no matter where APIs are developed or consumed."

API tools are evolving rapidly, and with third parties each having their own API interfaces, managing the API environment is still fraught with complexities and standards are difficult to establish and maintain (Longbottom 2015). IT leaders must therefore continue to monitor this marketplace and be prepared to evolve their tools and platforms as the API economy develops (Collins and Sisk 2015).

5. *Build APIs first.* "It is best practice to build the API before coding," said a member. The group agreed that APIs must be "rock solid" in design and connect to the common needs of the organization. Although the

services underneath them may evolve, an API defines the nature of the company's connection with its consumers and this shouldn't be taken lightly. Involving a broad spectrum of stakeholders and designing APIs based on future ecosystem requirements, rather than on existing infrastructure and data models, is also perceived as best practice (Malinervo et al. 2017).

6. ***Ensure control.*** Secure access to APIs is the foundation of the API economy. Without authentication and authorization, organizations are vulnerable in many ways. "We need to prove to our regulators that no one can see or manipulate our data," said a manager. "We need visibility about who is accessing it, how often, and where." Connections must therefore be certified and control has to be part of every change. With external APIs in particular, organizations need to develop secure ways to communicate and share. "There's much more rigor around this when APIs are involved," said a manager.

7. ***Connect APIs to business metrics.*** One of the most appealing features of APIs, according to the focus group, is the ability to track their usage. "This helps us to know what works," said a manager. "It also makes it much easier to measure business value." Once usage patterns are clear, it is then possible to develop meaningful business measures of an API's value, such as

number of users signing up for additional services, or capabilities accessed in new ways (O'Neill et al. 2017). "Linking APIs to business metrics helps us to focus on what APIs to develop and where to invest," said a member. "We should expect a business payback. If an API is not used, it's ineffective."

8. *Expose and address risks.* APIs create new levels and types of risk and business leaders are highly sensitive to these. "We must get these risks on our agendas. There are many and we are not adequately addressing them," said a manager. Some risks relate to bad data or poor decisions about data. Others relate to poor choices made by partners or about partners. And most important is reputational damage to core company products (Wixom 2016). Cyber-risks that can be exploited by hackers can be exposed when APIs are introduced (Collins and Sisk 2015). Finally, the effort involved to connect external APIs to existing systems is often underestimated (O'Neill 2016).

While these are the risks of which companies are most aware, there are other less obvious ones to address as well. The focus group noted that there is always a danger when an API is retired or a vendor goes out of business. If a company or its consumers depend on it, they may not be able to conduct their business. In some cases, these situations have led to lawsuits (Collins and Sisk 2015). API commercial agreements are often complex and poorly understood, again leading to potential legal liabilities. API pricing is a risk too. Subscriptions for APIs may start small and grow fast, leaving unprepared companies with sticker shock (O'Neill 2016).

Conclusion

Companies are increasingly seeking to connect with external third parties, whether software-as-a-service providers, partners, or app developers. At present, they are being very cautious about doing this through APIs because of the significant risks involved. However, as with other types of change, the focus group predicted that the risks will be addressed over time and disruption to existing business models will accelerate. "While it's an internal marketplace right now for us," said a manager, "we'll be out there in five years." In this chapter we have described how and why APIs are beginning to change IT and business, and ultimately our economy. We discussed the value APIs are expected to drive and presented a framework for developing API strategies. No one really knows how the API economy will shape up but "one thing we know is that we will look different ten years from now," the group concluded.

References

Anuff, E. "Almost Everyone is Doing the API Economy Wrong." *TechCrunch.com*, 2016. https://techcrunch.com/2016/02/21.

Clark, K. "Microservices, SOA, and APIs: Friends or Enemies?" *developerWorks*, IBM Corporation, January 21, 2016. https://www.ibm.com/developerWorks/.

Collins, G., and D. Sisk. "API Economy: From Systems to Business Services." *TechTrends 2015*, Deloitte Consulting.

Columbus, L. "2017 is Quickly Becoming the Year of the API Economy." *Forbes*, January 29, 2017. https://www.forbes.com.

Gilpin, M., and R. Marshall. "Reinventing Applications as Products for the Digital World." Gartner Research Report, ID: G003277399, May 2017.

Golluscio, E., A. Gupta, and M. O'Neill. "Design API Mediation Layer to Underpin your Digital Business Technology Platform." Gartner Research Report, ID: G00323828, May 5, 2017.

Graham, J. "Why Siri Won't Cooperate with Apps." *Toronto Star*, July 15, 2017, B14.

Hoelch, K., and P. Ballon. "Competitive Dynamics in the ICT Sector: Strategic Decisions in Platform Ecosystems." *Communications & Strategies* 99, Third Quarter, 2015, 51–70.

Isckia, T., and D. Lescop. "Strategizing in Platform-based Ecosystems: Leveraging Core Processes for Continuous Innovation." *Communications & Strategies* 99, Third Quarter, 2015, 91–111, 187, 189.

Longbottom, C. "The API Economy or the API Tower of Babel?" *ComputerWeekly.com*, August 12, 2015. http://www .computer-weekly.com/feature.

Malinverno, P. "The API Economy: Turning Your Business into a Platform (or your Platform into a Business." Gartner Research Report, ID: G00280448, February 19, 2016.

Malinverno, P., K. Moyer, M. O'Neill, and M. Gilpin. "Top 10 Things CIOs Need to Know about APIs and the API Economy." Gartner Research Report, ID: G0031885925, January 2017.

Narain, R., A. Merrill, and E. Lesser. "Evolution of the API Economy." IBM Corporation, 2016.

O'Neill, M. "Establish Governance of External APIs to Avoid Unpleasant Surprises." Gartner Research Report, ID: G00308763, July 22, 2016.

O'Neill, M., P. Malinverno, J. Herschmann, E. Golluscio, and D. Wan. "Create the Role of API Product Manager Part of Treating APIs as Products." Gartner Research Report, ID: G00320767, January 24, 2017.

Pettey, C. "Welcome to the API Economy." June 9, 2016. http://blogs.gartner.com/smarterwithgartner/author/cpettey/ (downloaded July 17, 2017).

Weill, P., and S. Woerner. "Thriving in an Increasingly Digital Ecosystem." *MIT Sloan Management Review* 56, no. 4 (Summer 2015).

Wixom, B. "Generating Business Value from Data." *Society for Information Management Advanced Practices Council*, presentation, May 3–4, 2016.

Chapter 3

Emerging Technologies

IT has struggled to get it right with emerging technologies (ETs). A perennial business complaint is that IT is not helping us see and implement the potential of new technologies fast enough. At the same time, there are also many cases where business has rejected IT requests for experimentation with new technology because it feels there are other things that will bring a higher and more immediate return on investment. But with countless opportunities available through emerging technologies, steps need to be taken in order to harness and manage this driver of innovation appropriately for a particular organization.

ETs are a big gamble for business. Investing in them can frequently mean failure—to deliver value, to be adopted, to be strategically significant. However *not* investing in them can mean falling behind, failing to be relevant to customers, losing market share, and having to continually play catch up in IT investment. Finding the sweet spot between these two poles and determining where and how to place bets on emerging technologies is an art, not a science. And it is frequently done poorly, both in business and in IT. As new technologies enter the marketplace at an ever-greater velocity, organizations more than ever need new ways to identify and assess emerging technologies, and to energize their organizations around imagining their possibilities.

There are at least four major components to effectively managing ETs (Weiss and Smith 2007; Fenn 2010). First, they must be identified. Second, they must be assessed for their business and technical potential. Third, potential technologies must be connected with real business needs and opportunities. And fourth, practices and skills must be in place to ensure that the right ETs are implemented at the right time.

Emerging Technologies in Business Today

The challenge of managing ETs is multi-dimensional and not limited to IT itself. Although it is common to speak of new or emerging technologies, what organizations really want is insights into how best to use technology in the marketplace

(Cusumano 2011). A significant majority of business executives now believe that technology can transform their businesses but they continue to be frustrated by the slow pace of change and how difficult it is to get great results (Fitzgerald 2014). Although this is not a new phenomenon (McKeen and Smith 1996), the pace of change for organizations has ramped up considerably in recent years. Today, companies in many industries are feeling increased pressure to find and develop innovative technology solutions that outpace those provided by their competition. Thus they are having to move faster and faster just to stay in the same place (Tiwana 2013).

Unfortunately, there is no "one size fits all" approach to addressing this challenge, said the focus group. The need for change and the pace of change depend on a number of factors, such as the market aggressiveness of the firm, the industry involved, risk and regulatory issues, and corporate philosophy (Sarner and Fouts 2013). Therefore, the group concluded that one of the most important questions for companies to ask themselves *before* determining how they want to manage ETs is: *Where do we want to be in the marketplace?* Some firms decide to be leading edge; others prefer to be fast followers; still others want to be in the middle of the pack. Within an organization itself, the appetite for incorporating ETs can also vary by function and between business and IT. "Our business units want to know: What will enable me to execute better, faster, or cheaper?" said one manager. "Our IT organization wants to know: What is the impact of new technologies on our governance, security, and data?"

Once this broad business context of firm readiness to integrate ETs is understood, it is important for an organization to establish an approach to making good decisions about ETs and how they will be used. ETs can be used to transform a business and gain and sustain competitive advantage but only if the strategic priorities of the organization are clear (Weiss and Smith 2007). Often, however, the vision for how to use ETs is unclear and unarticulated, leaving both business and IT frustrated and confused (Mangelsdorf 2012; Fitzgerald et al. 2014). In such cases, both groups are vulnerable to making inappropriate choices about ETs. The focus group noted that vendors may try to do an "end run" around IT principles and guidelines and attempt to exploit the business' frustration and ignorance, leaving an organization open to unexpected risks. On the other hand, IT can easily get caught up in new technology "hype" and overlook the business value such technologies should be achieving.

The focus group also pointed out the lack of clarity about what exactly an ET actually *is*. In some definitions, an ET is a technology that is not yet mature in the market; in others, it's any technology an organization isn't yet using. The group noted that their companies also distinguish between emerging *consumer* technologies and new *infrastructure* technologies. "We are much more flexible about adopting ETs on the periphery of our business," said one manager, "but we recognize that we need stability and a different approach to ETs with our core technologies." Overall, managing ETs is a bit like riding a tornado, the group concluded. Nevertheless, they recognized that their organizations need to better address ET management and develop some practices and principles for making good business and technical decisions about ETs.

Identifying Emerging Technologies

There is broad recognition in the technology community that it is not always easy to "know what you don't know." Thus the first step in better managing emerging technologies is to ensure that an organization has effective mechanisms to identify what technologies are available and how they might be used in their organization. For this reason most organizations use a variety of techniques to identify new and potentially useful technologies. These include:

- Vendor and industry conferences, events, and forums;

- White papers;

- Research and analysis boards such as Forrester and Gartner Group;

- Vendor and consultants' reports on future trends;

- Business partners;

- Research by central architecture groups.

The variety of these sources within individual organizations suggests that scanning for new technologies involves creating and tapping into an ecosystem of information offered by a broad variety of sources on an ongoing basis (Weiss and Smith 2007).

In addition, focus group members noted three other ways of identifying emerging technologies:

1. *Observing push technologies*—those that vendors are pushing or selling to create demand—and watching what is being used in the market, talking with peers in their industry or different industries, and addressing technology currency.

2. *Responding to pull technologies*—those that business functions or application development request to meet their specific needs.

3. *Screening for decentralized technologies*—those acquired by the business for their own specific purposes without reference to formal IT processes.

Altogether, this is a daunting task that is made even more difficult by the fact that each of the above types of information may be acquired by more than one IT group or individual. The focus group members noted that one of the biggest problems they had was a lack of communication between people doing this and other aspects of emerging technology work. Although most have a formal enterprise architecture group charged with developing a technology roadmap, the participants noted that such groups are often more removed from business needs than other parts of IT and have a mandate that includes broader infrastructure issues, such as incorporating legacy and upgrading existing technologies. Thus it is important to make managing ETs *someone's* job in the organization, although many may participate in the ET identification process.

Assessing Emerging Technologies

Although it is important to know what ETs are available, organizations have only a limited capacity to absorb them. Therefore it is critical to select only those few that will have the largest business impact. The focus group stressed that it is essential to thoroughly understand the business needs of the organization in order to make this selection. "This is something we need to do better," said one manager. "Our relationship managers are often too focused on more immediate matters and don't always take the time to explore future needs."

Assessment is all the more important because ETs are characterized by a low "signal:noise ratio," which tends to confuse both business and technology people about the potential of a new technology. "Signal" refers to indicators of value to a firm's core business, and "noise" refers to factoids, assertions, and beliefs about a technology that are *not* meaningful signals. "At the earlier stages of the life cycle of an emerging technology, the signal is faint and the noise is overwhelming.... A low signal:noise ratio means that information surrounding an ET is difficult to interpret, leads observers down multiple blind alleys, and requires ... effort and expense to discern meaningful insights" (Tiwana 2013).

Amplifying signals involves working closely with the business to better understand where and how value *could* be delivered with a new technology. One company's ET staff meet regularly with business leaders to ferret out opportunities by asking, "What do you wish you could do in your business if technology could be found to enable it?" (Weiss and Smith 2007). Several companies routinely hold internal briefings or events where selected ETs and their potential can be presented to executives. ET staff need a deep understanding not only of the business and its needs, goals, and strategies, but also of how the industry is developing because ETs can often provide firms with the opportunity to move into adjacent markets or develop products and services that are complementary to those that are already provided in an industry. Finally, business and ET staff can work together to amplify signals by applying different frameworks that challenge existing preconceptions and spot non-obvious applications. For example, each of a company's products and services could be assessed to determine if how they are purchased or delivered could be shifted from physical to digital. (For a more in-depth discussion of these frameworks see Tiwana 2013.) The goal of a preliminary assessment is to understand how an ET might affect the organization's products and services, work, new forms of external engagement, or business models.

Assessing an ET's technical potential involves a different set of lenses and is generally a more straightforward process. All focus group organizations have technology roadmaps that provide a baseline for determining many of the factors that could be critical to this assessment. These include: technical maturity levels; implications for integration, data, security, and operations; and complexity. The focus group also cautioned about reducing "noise," noting that much ET work is simply following the crowd, as opposed to true assessment.

Overall, assessment of any ET involves determining both relevance for the business and technical readiness. It requires having both a deep knowledge of the organization's business and a broad appreciation of technology. The focus group concluded that technology is meaningless unless it is understood in a business context, and they agreed that it is only through strong partnerships with the business that ETs can have the type of impact a business is looking for. "There are two places you can start to assess an ET," explained one manager. "You can look for an interesting technology that has a potential business benefit or you can find an interesting business opportunity and determine how technology can help you do business differently. Typically, you iterate between these two perspectives when assessing an ET, but you do it differently in every case."

Addressing Uncertainty

Once an ET has been deemed both relevant and technologically ready, there is an important further layer of evaluation that must be done, again most likely in several iterative steps. This is addressing the critical question of: *Should we be doing this?* ETs, more than other forms of IT work, by definition involve working with uncertainty. There are at least three types of uncertainty that need to be addressed before making a decision to move forward with an ET:

1. ***Market uncertainty.*** This is ambiguity about how the market will respond to an innovative new technology or application. One of the key aspects of determining this is assessing what complementary technologies are available that will make a particular innovation an attractive proposition (Tiwana 2013). Today, successful new technologies are unlikely to be stand-alone devices or applications that will single-handedly transform an organization or an industry. Instead, they tend to function with platforms, which act as the foundational technologies/products around which a broad ecosystem of other products and services are built by other companies (Cusumano 2011). When these are tied together, they create complementarities or combinatorial innovations that create new opportunities for business value (Tiwana 2013). Without such complementary products and services, an innovation often

fails. Unfortunately, the technology landscape is littered with such technologies that were innovative but ahead of their time because complements were not available. Without these, an organization can be too early to market with a particular technology. Furthermore, bundles of functionality are usually perceived as better value by users. Therefore, a crucial aspect of reducing market uncertainty is to identify the complements that will make an ET viable as an attractive proposition for consumers (Tiwana 2013). Market uncertainty can best be addressed by monitoring the ET ecosystem for the emergence of such complements (Fenn 2010).

2. *Technological uncertainty.* This is ambiguity about the maturity, complexity, or interoperability of any new technology. Many focus group organizations had sophisticated assessment processes for reducing this type of uncertainty. Tools can include: scorecards, radar screens, and watch lists. Enterprise architects have often driven the development of practices in this area. However, the focus group cautioned that because these groups tend to focus on core infrastructure and have more of a control mindset, there continue to be doubts about whether their organizations are investing in the right technologies for the capabilities they will need. "We need more of a formal incubation process," one manager said. Prototypes and "dabbling" with new technologies are key ways to deal with technical

uncertainty because they help develop a deeper understanding of an ET's strengths and weaknesses.

3. ***Economic uncertainty.*** Many focus group companies require a business case to proceed with the implementation of an ET, but setting too high a hurdle rate for these innovations will not mitigate uncertainty and will only result in a loss of innovation. Focus group members have found that a lower level of initial rigor is therefore required to prevent this. Also key is getting business sponsorship. "Unless someone wants it enough in the organization to act as a business sponsor, we don't go ahead with it," said a manager. "We go with the energy. We don't fight battles." Pilots can help address gaps in knowledge, determine how best to apply an ET in a live operational environment, and provide valuable information for a more complete business case. In one organization, after assessment has filtered out many ETs, pilots are undertaken with those remaining ones that have found sponsors. Of these, only about 10 to 20 percent actually move on to broader adoption (Weiss and Smith 2007). The key is to fail fast, said the focus group. From assessment to pilot to a decision about broader adoption should only take about three months. Developing mechanisms to achieve this is the best way to ensure that an organization can quickly adapt to changing market conditions (Cusumano 2011).

There is no way to eliminate all uncertainty with ETs, the group concluded, but the more items and people in the ET tornado, the greater the likelihood of getting the right technology in the right place at the right time. As with the assessment step, there is no structured process for addressing uncertainty so evaluating a new technology will typically iterate between each of the above items. Working through these activities will likely eliminate a larger number of technologies that are inappropriate for business, technical, or economic reasons, and provide a milieu that will enable better ideas to become clear.

Managing Emerging Technologies

Although none of the focus group members had a dedicated ET group, there was broad recognition of the need for more attention to ET management. "Most game changing organizations have an ET function," said one manager. Another added, "We need more resourcing for, engagement in, and processes for managing ETs more effectively. It may not be governance per se but we need to do a better job of facilitating it and coordinating knowledge." Many experts agree. An emerging technology process helps organizations become more selectively aggressive so they can counter the hype cycle and make strategic decisions (Fenn 2010). Organizations therefore need to build the skills, structure, and practices to determine which ETs will change the fundamentals of their business or industry and which can be ignored (Sarner and Fouts 2014).

Yet the focus group also pointed out that establishing effective ET management is a challenge. "We've tried three times to create an ET process and no one was motivated to participate," commented one manager. "We don't have any formal mechanisms to track or share information about ETs within IT," said another, "and we need this type of information sharing." The need for speed, the continuous hype and market pressures associated with ETs, existing work with a better business case, lack of sponsorship from business, and poor understanding of exactly *how* this rapidly evolving milieu might be managed successfully, all mitigate against developing a single set of processes for all ETs.

The group suggested that managing ETs for an organization's core infrastructure might need to be different from managing ETs for interaction with customers or specific communities. "We need a formal technology blueprint for our infrastructure," said a manager, "but we can be less formal with the periphery." This is a distinction that has been successfully used in a number of organizations. It suggests that governance for infrastructure is best centralized to enable synergies and scale economies, while decentralization is a better way to spot good business opportunities (Tiwana 2013). "We need the right engagement at the right time," explained a manager, "but it is less than governance and more like facilitation, leveraging knowledge, and coordination."

Organizations also need to develop new skills and capabilities for dealing with ETs because these are not the same as those needed in traditional IT. Focus group participants suggested that the following capabilities are needed:

• Strong business knowledge and ability to speak business language;

• Research skills;

• Visioning skills;

• Ability to partner with the business;

• An open mind, flexibility of approach, and comfort with uncertainty;

• Ability to explore alternatives and take advantage of serendipity;

• Relationship management skills ("but with teeth," said a manager) to work with different parts of the business and different vendors.

The group also stressed that any ET initiative should be designed to "fail fast" and respond rapidly to new information. ET projects are explorations to better understand business, technical, or economic questions. Therefore, the traditional Systems Development Lifecycle (SDLC) approach to IT work is inappropriate for ETs. "These projects are ideal for prototyping and agile methodologies," said a manager. However, once a decision has been made to implement an ET, there should be a formal technology transfer process that ensures an initiative is not adopted without a full business case and proper controls, and without addressing operational considerations.

Finally, the focus group stated that partners or vendors can play an important role in ET work. Typically, this relationship is most effective when helping companies identify new technologies and opportunities and in supporting the technology transfer process. Because of the significant business knowledge required in ET work, this is not an activity that can or should be outsourced (Fenn 2010). Furthermore, partitioning of ET work across more than one organization has been shown to decrease the likelihood of a successful outcome and increase the risk involved (Tiwana 2013). Therefore, companies seeking to develop effective ET management practices should be careful in how they involve vendors in this work.

Driving Opportunity with Emerging Technologies

As noted above, there is no single recommended process for managing ETs, and individual companies will approach it differently according to the needs of their business and industry conditions. Nevertheless, there are several general recommendations that can be made to organizations seeking to become more effective in their ET management:

1. *Make ET management someone's job.* Companies that are innovative with ETs have made it a priority and assigned resources to it. These *may* be in IT but may also be in a different organization that works closely with IT. In smaller companies, it may be just one person. However, it is important to note that the job is to coordinate, facilitate, and leverage ET knowledge, *not* to actually do all this work. Identifying, assessing, and evaluating ETs is a joint business–IT responsibility that should involve a number of people throughout the process.

2. *Always tie ET adoption to business value.* All too often, business or IT staff can fall prey to the hype of a spiffy new technology. ET management activities should be designed to help change mindsets in the organization about the possibilities a new technology enables, while never losing sight of the fact that it is the value a technology could provide that is the goal of any initiative.

3. *Educate others.* Many people need education to see the possibilities of a new technology. Prototypes, vendor visits, special events with thought leaders, and multidisciplinary innovation sessions are all helpful ways to encourage others to keep open and thoughtful minds.

4. *Go with the energy.* If a new technology doesn't attract a business sponsor, it should generally be shelved. In some cases, IT may have a budget for developing a few key projects further in order to demonstrate potential value but in short order someone in the business must be convinced enough of an ET's value to put some resources into it.

5. *Be brutal and quick.* With the market moving so quickly, an ET evaluation process must be designed to make rapid decisions. The goal of ET management should be to identify a wide range of potential technologies and then to rapidly filter them for business relevance, technical readiness, and economic viability and to focus on the very few that are going to deliver the most value to the organization. Short, speedy evaluation cycles will help to ensure the process doesn't get bogged down and that the organization is seen to be effective at evaluating and delivering on new technical capabilities.

6. *Don't downplay uncertainty.* Many companies downplay uncertainty either by setting a high hurdle rate for a new technology to be approved or by assuming that rigorous execution will mitigate it. The first approach can prevent promising new technologies from being adopted while the second can lead to the implementation of technologies that will not deliver the desired value. The key is to better understand whether

uncertainty comes from the market or the technology itself and to address it appropriately.

Conclusion

"Riding the ET tornado" is not for the faint of heart. It is both an art and a science to facilitate the right engagement of the right people at the right time. Effective ET management requires both deep business and broad technology skills to identify and evaluate the best technologies for the organization. The skills involved—ideation, agility, open-mindedness, and exploration—are not those in which IT is traditionally strong. Therefore, to further develop an ET capability, it will be important to seek out and encourage those with these skill sets. Although discovering ET opportunities may be "everyone's job," not everyone in IT will have the ability to assess them properly. This is likely why, in most organizations, IT has a poor track record when it comes to adopting new technologies. Therefore, even in the smallest of firms, if there is any desire to explore new technologies and the potential they hold for a business, someone with these types of skills must be given the mandate to facilitate and encourage these activities.

References

Cusumano, M. "How to Innovate when Platforms Won't Stop Moving." *MIT Sloan Management Review* 52 (4), Summer 2011.

Fenn, J. "Driving the STREET Process for Emerging Technology and Innovation Adoption." Gartner Research Report, ID: G0014060, March 30, 2010.

Fitzgerald, M., N. Kruschwitz, D. Bonnet, and M. Welch. "Embracing Digital Technology: A New Strategic Imperative." *MIT Sloan Management Review* 55 (2), Winter 2014.

Mangelsdorf, M. "What it Takes to be a Serial Innovator." *MIT Sloan Management Review* 53 (4), Summer 2012.

McKeen, J. D., and H. A. Smith. *Management Challenges in IS: Successful Strategies and Appropriate Action.* Chichester, England: John Wiley & Sons, 1996.

Sarner, A., and R. Fouts. "Agenda Overview for Emerging Marketing Technology and Trends 2014." Gartner Research Report, ID: G00255386, November 21, 2013.

Tiwana, A. "Separating Signal from Noise: Evaluating Emerging Technologies." *SIM Advanced Practices Council*, July 15, 2013.

Weiss, M., and H. Smith. "APC Forum: Leveraging Emerging Digital Technology at BP." *MIS Quarterly Executive* 6 (2) (2007).

Section II

Discovery Drivers

Once an opportunity is identified, effective management explores an innovation's potential value, enhances it iteratively, and situates opportunities within an overarching digital strategy. Chapter 4, "Thought Leadership," describes the missing links in thought leadership in IT and explores how to successfully foster new skills and capabilities in this area. Chapter 5, "Digital Strategy," examines the need for a distinct digital strategy that will drive an organization into the future. Bridging this gap effectively involves understanding the many dimensions of digital strategy, rethinking the organization's business model, and transforming IT processes to support its evolution. Chapter 6, "Experimentation," discusses the ways in which IT organizations often fail to learn from experiments and appreciate their missteps in particular.

Chapter 4

Thought Leadership

There is no question that both business and the IT industry are operating in times of unprecedented change characterized by considerable uncertainty (Mansharmani 2012; Kiron et al. 2015). On one hand, our increasingly connected and global economy and new business models driven by new digital technologies mean that businesses are finding it difficult to navigate the vague and poorly defined conditions in which they find themselves (Mansharmani 2012). On the other, new technologies are emerging all the time that could facilitate new business opportunities or undermine existing revenue streams. In the middle of all of this sits the organization's IT function, which is charged with assisting business leaders with determining where and how they can add value or change their business models to incorporate technology—often in fundamentally different ways than in the past. And on top of this, IT is being asked to radically change how it works to become more agile, effective, and faster, all while ensuring that the current day-to-day business of the organization flows smoothly!

Within this context, innovation can flounder without thought leadership. Thought leadership provides a steady hand in times of turmoil and helps close other innovation gaps. But where and how does thought leadership arise? "This is something we talk about constantly because the business wants it and IT wants to partner with the business," said one IT manager. "Business is looking to IT for ideas that will help make it better," said another, noting the value of a really good idea. But what really is a thought leader? And where does he or she fit into an IT organization? Can thought leaders be developed? Or are they born? IT organizations are wrestling with this concept and how it connects into their mandate to be innovative and agile digital strategists.

In this chapter we examine these questions, starting with a discussion of the nature of thought leadership and how it fits with IT's mandate. Following this, we describe some of the characteristics of an effective IT thought leader. Finally, we explore how IT functions can foster thought leadership successfully as well as the ways thought leadership can be inhibited in organizations.

What Is a Thought Leader?

There are many different and fuzzy conceptualizations of what makes a true thought leader (Prince and Rogers 2012). As a result, most people have no idea what the term means when it is used (Kim 2014). The focus group noted that thought leadership is not proactively defined in their organizations. "We define leadership and strategy, and we have a small innovation program, but we don't focus specifically on thought leadership," said one manager. Nevertheless, it is talked about so much in organizations that the term has actually been the subject of satirical columns that hold the "thought leader" up as "a new paragon to command our attention" (Brooks 2013).

What is clear is that thought leadership is a highly desirable characteristic both for organizations and for individuals to have at all levels. "We know that there is great power in good ideas," said a manager. "And that even little ideas can have great impact." But what does thought leadership look like in practice? There is no real consensus about this. The following is a short list of how thought leaders are described by the literature and the focus group:

• A thought leader is an authority in a specialized field whose expertise is sought and often rewarded (Wikipedia 2016).

- Thought leaders look for ways to make the organization better . . . there are no extrinsic rewards (focus group).

- Today's thought leaders can read complex business situations and bring out the best in other people. Thought leaders require personal traits that far exceed expert knowledge (Brendel 2016).

- Executives are increasingly defined by the degree to which they engage internal and external audiences with actions and ideas that are inspiring and selfless, and that extend well beyond the scope of their core mandate (Kim 2014).

- Thought leaders can be anywhere in the organization. Being in a position of power does not make one a thought leader (focus group).

- Thought leaders are strategic thinkers. They nurture relationships and social networks, asking good questions of others, and synthesizing that information into actionable intelligence (Brendel 2016).

- Thought leaders look for inflection points. They educate and create awareness. They give away their ideas to be operationalized (focus group).

- Thought leaders change the way people think and what they do. The best ones are actually trying to address a problem at hand and not just talk about it (Kim 2014).

- Real thought leadership tends to be disruptive and uncomfortable (focus group).

- Thought leaders must be willing to buck the status quo (Brosseau 2016).

- Thought leadership can come from group dynamics and discussion (focus group).

- Thought leadership stems from passion (focus group, Brosseau 2016; Kim 2014).

Connecting all these dots into a coherent picture of a thought leader can be frustrating, but some key themes emerge from these statements. It appears that thought leadership can be exercised in many different contexts in both business and IT. It is also clear that it relates to an idea to make things better or do things differently, and this often challenges the status quo as a result. In addition, it takes more than a good idea to provide thought leadership, though good ideas are where it starts. Thought leaders need to have knowledge, credibility, and persuasiveness to bring their ideas to a point where they'll get traction, even if they don't implement them themselves. In short, "While we really don't know how to get thought leadership, we know it when we see it," concluded a focus group manager.

Thought Leadership and IT

Thought leadership is an important component of IT work at all levels, said the focus group, whether or not it has a formal mandate to do it. "We must accept that the business wants thought leadership from us," said a manager. This can be accomplished in different ways and at different levels, according to the context involved. "A thought leader can be in any role," said one manager. "He or she could be a technology leader, have deep business knowledge, provide horizontal leadership across company silos, or be in a formal position to provide long-term leadership. Thought leadership can also be born at different stages of an initiative." The common theme is the *idea*.

Innovation and thought leadership work hand-in-hand, the group agreed. In another focus group, members explained how IT ideas work in innovation:

Innovation . . . refers to the process whereby a company creates new things that deliver value . . . The first stage is ideation—generating innovative ideas. There are many ways of doing this, ranging from focused executive meetings to the modern online version of the suggestion box. Ideation addresses two questions: How do we get people to share their ideas? How do we respond to their ideas? In most cases, the focus group has found that there are lots of ideas out there. Attempts to stimulate innovation led to an initial deluge of new ideas. However, with very little ability to screen and prioritize or act on them, the ideas soon dried up ... [because] the biggest reason why people do not share their ideas is that past experience has shown them that management doesn't respond to or act on them.

But thought leadership is also different from innovation. "Thought leaders *spark the innovation cycle*," a manager explained. "They provide the connection, the synthesis, and the intellectual energy to start down a different path."

Thought leadership in IT also has a strong technology connection, although "we can't think about technology too narrowly," said a manager. As we have noted elsewhere:

A perennial business complaint is that IT is not helping it see and implement the potential of new technologies fast enough. At the same time, there are also many cases where business has rejected IT requests for experimentation with new technology because it feels *there are other things that will bring a higher and more immediate return on investment.* (Smith and McKeen 2013b)

This highlights an important distinction between thought leadership and managing emerging technologies. Thought leaders "*energize* their organizations around imagining their possibilities" (Smith and McKeen 2014). Thought leaders must not only understand technical possibilities, they must "raise the game," "motivate others to see them as well," and "work with groups to change their mindset and collaborate to deliver on the new idea," said the focus group. "It's a change agent role."

Although it is more common for IT leaders to speak of new or emerging technologies, what organizations really want is insights into how best to *apply technology* in the marketplace (Cusumano 2011). "Our business units want to know: What will enable me to execute better, faster, or cheaper?" said one manager. Often, however, the vision for how to use emerging technologies is unclear and unarticulated, leaving both business and IT frustrated and confused (Mangelsdorf 2012; Fitzgerald et al. 2014). In such cases both groups are vulnerable to making inappropriate choices.

This underscores a third important distinguishing element of thought leadership. It is a *leadership* role requiring a number of soft skills that are now essential at all levels of the IT organization, chief of which is communication. Elsewhere, we have noted the importance of developing general leadership skills further down in IT than in other parts of the organization:

> There is no question that individuals within IT have more opportunities to affect an organization, both positively and negatively, than others at similar levels in the business. This fact alone makes it extremely important that IT staff have much stronger

organizational perspectives, decision-making skills, entrepreneurialism, and risk assessment capabilities at lower levels. Today, because even small decisions in IT can have a major impact on an organization, it is essential that a CIO be confident that his/her most junior staff have the judgment and skills to take appropriate actions. (McKeen and Smith 2012)

The focus group stressed that although formal IT leaders are not necessarily thought leaders, leadership is implied in the concept of thought leadership. These distinguishing features of thought leadership work together to deliver *business impact* (see box on next page). One manager described an initiative in his organization that brought each of these dimensions of thought leadership into it (identified in brackets):

We implemented an application design change [sparks new idea] through relocation of mainframe processes [applies technology to make new things possible]. This was enabled by a group of architects who had a wider vision of systems and architecture [energizes others] and motivated by their desire to deliver on their mandate [leadership in context]. This resulted in one million fewer batch jobs each year [delivers business impact].

Finally, the group stressed that it is important for organizations not to rely on external providers for thought leadership. "Most external providers don't know the dots to connect," said one manager. "They compare us with organizations that look like us. This can lead to a self-affirming ecosystem that believes 'we're good.'" Internal thought leaders pay attention to what's happening in much wider contexts outside their organizations, such as academia, research organizations, and other sources beyond their industry. They also scan the market for new ideas and emerging technologies that will enable them to lead cross-organization and internal ideation and innovation teams to leverage their business model (Burton et al. 2016).

In summary, thought leadership in IT is a central part of the innovation process both within IT and in the business. It challenges current ways of thinking and working in an organization by seeking out new ideas, new technologies, and new ways of putting processes and products together and then *leading* the appropriate groups to help them see

Distinguishing Features of an IT Thought Leader

- Sparks innovation with new ideas;

- Energizes others around new possibilities;

- Applies technology for value;

- Leads in context;

- Delivers business impact.

the real possibilities for value in this new approach. Figure 4.1 summarizes the place of thought leadership within the organization. It shows that thought leadership sits squarely at the center of current business and technical practices and new business and technical opportunities, and serves as the starting point for innovation, nurturing new ideas until they gain support in the organization. This is a challenging, often thankless task, but one that gets at the very heart of the role business wants IT to play in today's complex and uncertain world.

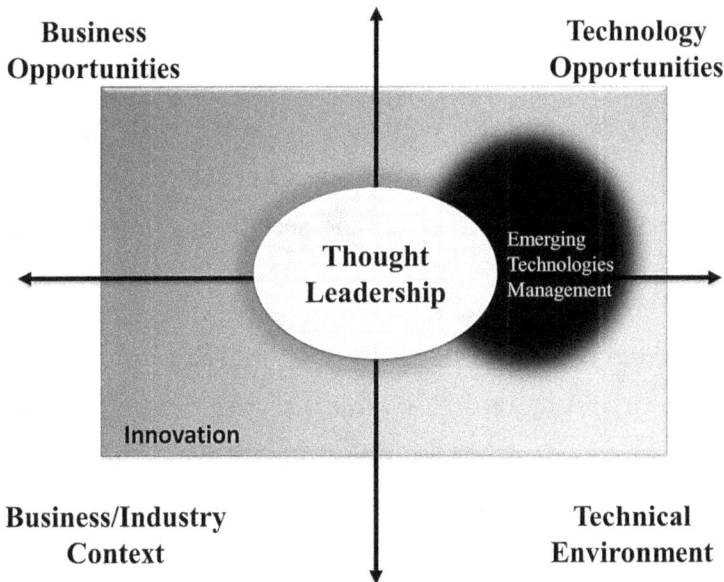

Figure 4.1 Thought leadership is central to IT's innovation responsibilities.

Characteristics of an IT Thought Leader

Although the group had difficulty defining thought leadership in general terms, it was very clear about the characteristics of a good IT thought leader. "We may not know exactly what a thought leader is," said a focus group participant, "but we know one when we see one." Although the ideas a thought leader comes up with are important, the skills that he or she uses to sell them and the ability to synthesize his/her ideas with those of others are essential for a successful thought leader, said the group. Some of the key characteristics of a thought leader in IT are:

- *Idea-generator.* A thought leader is forward thinking and adaptive, offering intellectually attractive ideas. He/she thinks outside the box and sees things in ways that others don't.

- *Visionary.* A thought leader is more than just an idea creator and more than a subject matter expert. She has a natural curiosity and insight into a situation and takes an idea to the next level, working it through, and seeing patterns. Because of this, a thought leader sees a path forward when others can't.

- *Consultant.* A thought leader helps others think through ideas and looks for ways to add value. He recognizes the importance of *selling* the idea to others, not pulling control levers.

- **Synthesizer.** A thought leader connects the dots between their idea and other people's ideas to provide a more compelling vision.

- **Collaborator.** A thought leader listens to others and adjusts her ideas based on their input, rather than clinging to an original idea. She recognizes that thought leadership is a process of leading the refinement of an idea and that teamwork can make good ideas better. A thought leader is therefore happy to share the credit for a successful idea.

- **Courageous.** A thought leader believes in his idea and challenges others or disrupts established ways if necessary. "He is sometimes a lightning rod for change," said a manager. "He questions how things work and puts forth ideas that do not align with the views held by others."

- **Persistent.** A thought leader understands that unconventional ideas are rarely, if ever, accepted immediately. She recognizes that influencing others to follow a new path is invariably a process of communication with many people over an extended period of time.

- **Communicator.** Thought leaders recognize that selling their ideas is a must. "This involves tuning into the zeitgeist of the situation," said a manager. They package ideas effectively and simplify complex situations, articulating ideas in ways that resonate with others, making them want to move ahead. In this regard, storytelling is a very useful skill.

- *Credible.* A thought leader has credibility with the subject area involved, either in the business or in the technical function. As a result of this credibility, people follow thought leaders because they believe in what they're selling.

- *Action-oriented.* While not necessarily a formally appointed leader, a thought leader drives action related to an idea and leads others to deliver a useful outcome. As a result, thought leaders need some implementation skills to understand and assist the delivery process.

- *Self-motivated.* A thought leader works for intrinsic value and the opportunity to make a bigger difference.

See the box on the following page for how these characteristics come together in a real thought leadership example.

Thought Leadership in an IT Function

"One of our enterprise architects became convinced [self-motivated] that DevOps would be valuable in her organization [idea generator, visionary]. She read widely [credible], wrote papers about it [synthesizer] that she passed around to key IT managers [courageous, communicator]. Slowly [persistent], she encouraged people to discuss what might be involved [collaborator, consultant], and quietly supported pilot efforts to explore its value [action-oriented]." (Focus group member)

Driving Discovery with Thought Leadership

Improving thought leadership in IT involves not only accessing and utilizing people's ideas but also creating an environment in which thought leadership is nurtured. Clear communication about the company's strategy and end goals from its leaders is a broad prerequisite or else thought leaders might develop disruptive ideas that don't add value (Fenn and Messaglio 2016). Thought leaders come in three types: natural thought leaders who don't have to be nurtured; one-hit wonders who may have a single good idea; and thought leaders who can be developed and taught over time. Each of these can benefit from a number of proactive steps IT leaders can take to foster leadership skills in their departments.

The first step is to harvest the ideas that people have, concluded the focus group. Many companies have already introduced some activities to collect and assess ideas from a broad spectrum of employees and others. IBM's "Innovation Jams" in which employees and ultimately vendors, partners, and spouses were invited to contribute ideas were among the first to formalize this process.[1] Other firms have introduced innovation contests (Smith and McKeen 2013a) or hackathons, where individuals or groups are invited to submit ideas, which are refined by the crowd, and some selected for a proposal to business leaders and prototyping. These are designed to surface unrecognized opportunities. Other ways of doing this are through workshops to help envision a future state or business model and by introducing an internal venture capital model (Applegate and Meyer 2016; Fenn and Mesaglio 2016). Effective processes collect, refine, and connect good ideas to raise the game for the organization. One of the least effective ways to do this, said the participants, is to create an elite innovation group, because this tends to discourage contributions from non-group members. However a company chooses to collect new ideas, it is imperative that they be seen to utilize them. "Organizations not only need great ideas, they need to deliver on them," said one manager. Failure to do so is the single biggest reason why people do not participate in

---

[1] https://www.collaborationjam.com/.

idea-generating activities, said the focus group. "You must take these ideas seriously," said another manager. "There must be evidence from senior management that something is happening with them."

Some companies have also changed their hiring and staff evaluation procedures to reflect their desire for more thought leadership skills. "Innovation is now one of our nine staff competencies," said one manager. "We are hiring for thought leadership aptitude," said another. "We have implemented an annual achievers innovation award," said a third. The key skills they are looking for, in addition to those mentioned above, include intellectual curiosity, empathy, broad experience with deep skills in one or more areas; and the ability to synthesize and apply different perspectives in a variety of business contexts (Morelli and Holub 2008). These people may be currently working as or aspiring to be enterprise architects or relationship managers whose roles naturally cross internal and external boundaries (Mok and Berry 2016).

Thought leadership can also be fostered both at an individual and a group level. Some focus group companies have formal technology leadership programs for new talent with high potential. These provide several rotations in different parts of the organization to give them cross training and a broader perspective before settling them in a particular area. Others provide coaching for high potential talent. "We provide opportunities for our best people to work on prominent projects," said a manager. "This empowers them to stand out."

Another company encourages IT staff to visit the business to find out how users really use the technology currently provided. "If you think you understand how they're actually using it, you're wrong. These visits are always enlightening," said another. Use of multidisciplinary teams creates opportunities to build relationships and stimulate ideas. These blend individuals with creative skills, user experience, engineering background, and functional knowledge to foster, evaluate, and implement new ideas. They also help build cross-functional relationships that can be useful in the future.

In addition, companies need to create a positive environment where thoughts and leadership can flourish at all levels. This is probably the biggest challenge for IT management (McKeen and Smith 2012). Unfortunately, organizations often take creative people and put them in soul- and idea-crushing environments (Fenn and Messaglio 2016). "We need to have a general culture where people feel free to express their ideas," said a manager. "This involves good, open leadership at all levels." Companies and leaders also need to be able to listen for and act on new ideas. "We train our leaders how to dialogue with their direct reports to promote a positive environment," said another.

A big challenge is to guard against "big egos" in an organization—those who dismiss or appropriate others' ideas. And some groups can be threatened by disruptive ideas—particularly if they are not part of that group—and fail to follow up on them. "Some parts of our company don't want thought leadership," said a manager. "We need to find a good fit for our thought leaders or they will be frustrated."

Finally, companies must ensure that staff have the *time* at all levels to work as thought leaders. Work stress, information overload, and multitasking can all inhibit the processes used for reasoning and forward thinking (Webb 2016). "Organizations must give people permission to spend time on these activities," said a manager. "Too often we're too busy and swamped with everyday work to do this." Leaders must promote taking breaks and going offline for important thinking to foster thought leadership. They should also encourage staff development in non-traditional subjects and foster connectivity and education (Bannister et al. 2015; Webb 2016).

Conclusion

Thought leadership is what business wants from IT. That said, it can often be a struggle to deliver. Traditional, hierarchical, siloed organizations tend to inhibit the type of creative, cross-functional thinking needed in thought leadership. Failure to make thought leadership a management priority can mean that organizations lack the processes and culture needed to foster it at all levels. Task-oriented management and pressure to deliver on a day-to-day basis can squelch innovative ideas. IT is moving into a brave new world where creativity, innovation, and new ways to deliver value are the norm. It is therefore imperative that IT adopt new ways to foster this type of thinking by creating opportunities to generate and harvest new ideas, promoting ways to develop and encourage thought leaders, and ensuring an encouraging culture that provides positive feedback, opportunities to develop cross-functional relationships and knowledge, and the time to engage in forward thinking and idea synthesis.

References

Applegate, L., and A. Meyer. "Launching 1871: An Entrepreneurial Ecosystem Hub." *Harvard Business School Case N1-816-090*, April 20, 2016.

Bannister, C., J. Pennington, and J. Stefanchik. "IT Worker of the Future." *Deloitte Tech Trends 2015*.

Brendel, D. "It's Time to Revisit the Concept of Thought Leadership." *Huffington Post*, May 16, 2016. http://www .huffingtonpost.com/david-brendel/its-time-to-revisit-the -c_b_9979600.html.

Brooks, D. "The Thought Leader." *New York Times*, December 16, 2013.

Brosseau, D. "What is a thought leader? FAQ." *Thought Leadership Lab*. http://www.thoughtleadershiplab.com/ Resources/WhatIsaThoughtLeader (accessed August 2, 2016).

Burton, B., M. Blosch, J. Dixon, and C. Howard. "Define Your Future Role as a CTO in the Digital Age." Gartner Research Report, ID: G00312950, July 20, 2016.

Cusumano, M. "How to Innovate When Platforms Won't Stop Moving." *MIT Sloan Management Review* 52, no. 4 (Summer 2011).

Fenn, J., and M. Messaglio. "Drive a Creative Culture Through Activities, Education and Attitude." Gartner Research Report, ID: G00304117, June 10, 2016.

Fitzgerald, M., N. Kruschwitz, D. Bonnet, and M. Welch. "Embracing Digital Technology: A New Strategic Imperative." *MIT Sloan Management Review* 55, vol. 2 (Winter 2014).

Kane, G., D. Palmer, A. N. Phillips, and D. Kiron. "Is Your Business Really Ready for a Digital Future?" *MIT Sloan Management Review*, July 2015.

Kim, C. "Think You're a Thought Leader?" *Financial Post*, March 7, 2014.

Mangelsdorf, M. "What it Takes to be a Serial Innovator." *MIT Sloan Management Review* 53, no. 4 (Summer 2012).

Mansharamani, V. "All Hail the Generalist." *Harvard Business Review*, June 4, 2012.

McKeen, J. D., and H. A. Smith. *IT Strategy: Issues and Practices (2nd ed.)*. Upper Saddle River, NJ: Pearson Prentice-Hall Education, 2012.

McKeen, J. D., and H. A. Smith. *IT Strategy: Issues and Practices (3rd ed.)*. Upper Saddle River, NJ: Pearson Prentice-Hall Education, 2015.

Mok, L., and D. Berry. "Ten Absolute Truths about Talent Management in Digital Business." Gartner Research Report, ID: G00303502, July 11, 2016.

Morelli, D., and E. Holub. "Inside the Concept of Versatilists: What are They and How do CIOs Develop Them?" Gartner Research Report, ID: G00159537, September 9, 2008.

Prince, R., and B. Rogers. "What is a Thought Leader?" *Forbes*, March 16, 2012.

Smith, H. A., and J. D. McKeen (a). "Organic Innovation at EMC." https://www.ciobrief.ca, 2013.

Smith, H. A., and J. D. McKeen (b). "Innovation with Technology." https://www.itmgmtforum.ca, 2013.

Smith, H. A., and J. D. McKeen (c). "Managing Emerging Technologies." https://www.itmgmtforum.ca, 2013.

Webb, C. "How Small Shifts in Leadership Can Transform Your Team Dynamic." *McKinsey Quarterly*, February 2016.

Wikipedia. "Thought Leader." https://en.wikipedia.org/wiki/thought_leader (accessed August 2, 2016).

Chapter 5

Digital Strategy

The word "digital" appears everywhere today, and there is a growing consensus that the world as we know it is in transition from our "old" ways of doing business and using technology to "new" ways of thinking about, living with, and working with technology. "The change is different in scale because it's also a disruptive change in how society works," said one IT manager. "There's a lot of hype about digital but there's also a lot at stake." "We've seen an unbelievable shift in the past five years," said another, "and in the past six months, 'digital' is a given."

There are a lot of reasons why organizations are feeling a sense of urgency about "going digital" and the pressure to change. Buying patterns are evolving. Companies are seeking to differentiate themselves with consumers (Baggi 2014; Hirt and Willmott 2014; Press 2016a). The cost of delivering IT solutions is changing. What would have been very expensive and time-consuming to develop in the past can now be delivered in just weeks or months by combining apps and off-the-shelf and cloud-based solutions (Willmott 2014).

New competitors can now attack specific areas of a company's value chain without needing to develop the whole thing themselves (Digital Strategy Conference 2015). This lowers barriers to entry and enables entrepreneurs to cherry-pick subcategories of products and undercut pricing (Hirt and Willmott 2014). In turn, these factors are changing the competitive landscape, undermining established business models, and putting pressure on business and IT alike to change rapidly. "There's a real sense of fear in our organizations that it can't be business as usual and concern that we won't be able to keep up," said one manager. It is like a perfect storm where organizations realize the imperative to "go digital," fear that competitors will get there first, and wonder how, where, and when to make the transition. Discovering how to drive innovation through the successful articulation of a digital strategy provides a path forward.

These elements have already profoundly changed the strategic context of business (Hirt and Willmott 2014), and this is driving companies to look at more holistic ways to respond to it. "We've had lots of individual strategies in the past but never an overarching one," said one manager. "In this environment, most strategies only have a shelf life of about five minutes." Nevertheless, companies are pushing forward to deal more comprehensively with the inevitable changes that are coming. Most are still struggling to develop a coherent digital strategy, while about a quarter are lagging in this regard, and another quarter are more mature (Kane et al. 2015b). All recognize that the move to digital will place a huge burden on IT to change. One study estimated that by 2018, 35 percent of IT resources will be spent on supporting the creation of new digital revenue streams and that this will grow to 50 percent by 2020 (Press 2016a). IT staff skills will also need to change to incorporate new technologies, new ways of working, a new focus on data, and an increased emphasis on risk, security, and compliance (Press 2016a). Thus, creating a successful digital strategy and an organization to support it is now a top priority for both business and IT management (Iyengar and Mok 2014).

In this chapter we first examine the meaning of the term "digital strategy." Then we explore the value of having a digital strategy as part of a company's business and IT strategies. We describe how to go about developing a digital strategy. Finally, we look at the broader implications of digitization for the organization and what further changes will be needed to support a successful strategy in an enterprise.

What Is a Digital Strategy?

"Few terms have created as much confusion or have as many completely different definitions as a digital strategy," wrote one researcher (McGee 2015). In the focus group everyone agreed that there is no commonly understood definition of the concept. Some members pointed out that "everything is digital," so it's just an evolutionary development of traditional IT strategy. Others strongly disagreed. In the end, all recognized that digital strategy is an evolving concept and that "where you're going, depends on your industry and where you're starting from." Researchers corroborate this and suggest that "depending on your sector and industry, the perception and role of digital changes radically" (Digital Strategy Conference 2015).

Regardless of the starting point, there is broad acceptance that digital strategy is different from traditional business and IT strategy in several key areas:

- *The unknown.* Businesses are changing dramatically in very short periods of time as a result of applying new technologies to traditional business models. Unlike traditional strategies, which were longer term and based on well-established business models and competitive landscapes, digital strategy is dynamic, with business and IT leaders working together to guess what the future will be and revise their strategies on the fly (Kane et al. 2015a).

- *Complexity.* Digital strategy can involve every part of the organization—functions, processes, products, services, data, technology, and employees—as well as partners, suppliers, customers, the value chain, and competitors. It seeks "methods to employ information and IT in new ways to create new value with new products, services, business practices, or even new business models" (McGee 2015).

- *Holistic.* Companies usually start off with isolated digital initiatives, but as their understanding and capabilities mature, they realize that the true value of a digital strategy lies in its ability to present a coherent and integrated approach to connecting people, processes, and things, and to enable them to communicate with each other via the Internet in ways that have never been possible before (Lopez 2015).

- *Exploration.* Unlike traditional strategy that seeks certainty about what to do, digital strategy stresses innovation, experimentation, and exploration. This requires different ways of thinking and working that incorporate both new ideas and tools with a supportive and flexible IT infrastructure and a disciplined and practical approach to development, implementation, and evaluation (Press 2016b).

- *Scope.* Although traditional business and IT strategies were designed to bring about change, they were always focused within a traditional business model. Mature digital strategies, on the other hand, seek business transformation by incorporating three key trends—big data and analytics, the internet of things, and artificial intelligence—to create new business designs that connect and/or integrate business assets (people, processes, and things) beyond the IT function and beyond the control of any one company (Press 2016a; Lopez 2015).

In short, a digital strategy is only the first step in a new approach to using business and technology. It outlines a map of what a company wants to become. Simply deploying a "digital something" will not automatically accomplish this or change mindsets, cultures, or work practices. Therefore, a company needs to be able to clearly articulate what it wants to accomplish and how it wants to operate as a digital business (Press 2016a, Baggi 2014). Failing to do this is a major barrier to progress and is now a concern reaching up to board level in organizations (Kiron et al. 2013b) (see box).

After reflecting on how these dimensions change the nature of strategy, the focus group came up with the following definition:

A digital strategy is a means of embracing new and different technologies in ways that challenge operational and value assumptions and which integrate them with existing technologies to deliver new products, services, business models, revenue streams and/or customer/stakeholder experiences.

The Importance of Digital Strategy*

"In a stunning decision [a Toronto-based retailer] replaced its CEO with [a person] best suited to deal with the myriad challenges of a rapidly changing, uncertain retail environment . . . [The] board was concerned the company's digital retail strategy was inadequate in the face of fast-moving competitors . . . The company's immediate priorities are finding and implementing the right strategy . . . Every day customers are demanding more control of their shopping experience . . . [The board is concerned about] . . . 1) the appropriate digital strategy, 2) the pace of its implementation, [and] 3) the role of acquisitions in implementing the company's digital strategy. . . . The change in the CEO reflect[s] that traditional retailing paradigms are under stress and must evolve."

* *Globe and Mail,* July 14, 2016

What Is the Value of a Digital Strategy?

A digital strategy serves as a tool for communicating these concepts to the organization and answers the following strategic questions:

1. Where will we choose to play?

2. How will we win?

3. What are the unconventional insights that matter?

4. What new concepts can we design and test?

5. What experiences do we want to create?

It helps both business and IT leaders to think about the true value of digital early on and identify both new opportunities and potential threats, so they can come up with the right balance of attack and defend positions (Willmott 2014).

In addition, there are a number of other benefits to having a well-thought-out and clearly communicated digital strategy, such as:

- *A mechanism for thinking about new directions for business.* The focus group felt that, if nothing else, the hype about and pressure to become more digital would be a catalyst for IT and business leaders to re-evaluate all aspects of their business model, from revenue streams, products, services, and channels to more mundane matters such as new practices and ways of working to save money, improve productivity, and create value.

- *An overarching view of multichannel business.* Focus group members stressed that old channels don't go away in this new digital world. New channels such as mobile and sensors supplement existing physical, telecommunications, and web channels. Ideally, together these create a richer experience where the channels complement and reinforce each other. However, if channel strategies are developed and implemented separately by different organizational silos, this opportunity will be lost.

- *The ability to mutually reinforce the physical and digital experience.* One company with a more advanced digital strategy found that although online is very important, bricks and mortar are not going away. "It's about both," said the manager. "There has to be a shared e-commerce model because physical touch is still important for much shopping. There is a lot of power in the combination of digital and physical. The proof is in the sales figures," he said.

- *An organizing metaphor for change.* Whereas in the past, IT work was grouped into projects, a digital strategy provides a more comprehensive roadmap for change with technology. It also makes it easier to track and adapt to changes in a particular industry and clarifies digital's impact on a business.

- *Improved connection with internal and external audiences.* A digital strategy ensures that a customer's overall experience with a company and all forms of customer engagement are considered by leaders, according to the focus group. Digital channels can also be used to enhance interactions with suppliers, stakeholders, and employees, and lower the cost of transactions. "This connectivity highlights the fact that a digital strategy is a two-way concept," said a manager. "With it, companies can now learn from their customers and others and even from their products in real time."

- *Improved integration.* When all the pieces in a process or an experience work together in a holistic way, companies can accomplish amazing things. Done well, this results in better leveraging of company assets (Kane et al. 2015b).

- *Improved innovation.* With a clear vision in place, innovation can be more focused and guided by both digital strategy and ongoing feedback. This facilitates continuous improvement and the development of new business and revenue models.

- *Improved business decision-making.* Real-time feedback from customers, suppliers, and employees, as well as from the big data collected by various types of digital technologies enables leaders to make improved decisions and reduce operational risk.

- *Improved employee engagement.* One unanticipated side-effect of the implementation of a successful digital strategy is a positive cultural shift in which employees feel more connected to the company, engaged with its brand, and aware of its strategy (Kane et al. 2015b).

"Our digital strategy will evolve," said an IT manager, "but we are excited that it will help to shape new directions for our business in ways that we haven't been able to achieve before." A digital strategy therefore represents a new way for business and IT leaders to work together to achieve new types of value with new technologies in new ways.

Developing and Implementing a Digital Strategy

As noted above, developing and implementing a digital strategy is a continuous process of ideation, experimentation, and evaluation. "It's a journey, not a destination" (Shen 2015). Simply providing some tools and praying that they will be used is not a recipe for success (Kiron et al. 2013b). Although it is true that digital strategy is akin to guessing at the future, there are ways to guide and structure these guesses. Developing and implementing a digital strategy takes both vision and discipline, said the focus group. They identified several practices that can assist leaders in narrowing and focusing their efforts:

- *Seek broad community engagement.* It is important to recognize that ideas and complementary skills can come from anywhere in the world (Press 2016a, Baggi 2014). Advice and input should be sought from leaders in a wide variety of industries (Baggi 2014). In some cases, organizations that have been competitors can potentially be collaborators (Willmott 2014). In addition, IT leaders must seek to ensure their staff is thoroughly engaged with the business and also with vendors, partners, and even customers. "A big challenge for IT leaders is that, while they have a good view of the overall organization, they are not experts in it," said one manager. "They should therefore explore where the knowledge lies in their organizations and plant seeds, even if they don't get the credit for the new ideas. It's a pioneering role."

- *Think backwards and then focus.* Since the future is unknown, organizations have the opportunity to create it for themselves. Experts recommend starting with a long-term vision of where the business needs to go and then reverse engineering from there, identifying the capabilities needed and then setting priorities (Kane et al. 2015a; Kane 2015). Following this, a small number of focused first steps can be identified. "Start small," one manager advised. "Set the stage and then help move change along with small milestones."

- *Nurture new ideas.* Companies need to cultivate new ideas in a systematic way. There are many ways to do this, such as internal crowdsourcing, innovation jams, or establishing a process for identifying and evaluating new ideas. However a company chooses to do this, it is most important that its leaders endorse and model their support for new approaches, experiments, and collaborative processes. In addition, companies must understand the innovation life cycle, its role, key transition points, and the different types of people needed at different points in the cycle, said the group.

- *Consider the role of the CIO.* Although some companies are creating Chief Digital Officers to lead the development and implementation of digital strategy, others believe this is the CIO's responsibility. "CIOs should be leading innovation because technology is driving serious change in our organizations," said one manager. Regardless of who leads, all senior leaders need to be actively involved, including the C-suite, line of business heads, the board, and finance (Kiron et al. 2013b). Ideally, the CIO should aim to catalyze innovation in the business, said the focus group, by asking business leaders, "Have you thought about this?"

  CIOs are also best equipped to understand the cutting edge, said the group. They can bring innovators and opportunities into the organization and then translate these into organizational language. CIOs and other IT leaders should therefore aim to be connectors, catalysts, pioneers, and incubators of change. In turn, they will need to be supported by two sets of people in IT: idea generators and solution generators.

- *Communicate constantly.* As with any major change, leaders need to invest time and effort in explaining what they are planning to do and what new ways of working will be involved and expected (Kane et al. 2015a; Kiron et al. 2015b). A digital strategy needs to be clearly communicated to all levels of the organization and all stakeholders. The better leaders can communicate their strategy, the more engaged the staff will be (Kane et al. 2015a,b).

- *Establish a digital business structure.* There is no "right" way to do this, said the group, but there is widespread agreement that however digital initiatives are structured, they need to be composed of cross-functional teams at all levels. Some companies are creating separate digital organizations, sometimes in separate premises, in order to break away from traditional cultures and work practices. Others place it in the marketing organization since many digital initiatives are customer-facing. Still others centralize strategy and decision-making while distributing teams where most appropriate (Kiron et al. 2013a,b; Hirt and Willmott 2014). The most important elements of any structure are an agreed-on strategy, involvement of both business and IT, strong support for new ways of working, and continual evaluation of future directions.

A mature digital strategy will also include:

1. *A digital experience design.* This outlines a long-term vision for the customer experience, including estimates of benefits, customer personas, a customer story map, app features, wireframes of key processes, and detailed user stories.

2. *A digital operating model.* This assesses the options, criteria, and path that will assist selection of the technology required. It includes evaluation of the current architecture and required web services, a list of integrations required, and proposed solution architecture.

3. *A digital platform assessment.* This outlines the timing, effort, and costs required for the organization to deliver multiple releases.

4. *Delivery plans for specific solutions.* These plans should include: creative concepts and designs, developer-ready requirements, prototypes for testing, deployed mobile apps, change and communication plans and support, and assistance in promoting and marketing apps and determining and measuring business value.

Driving Discovery with Digital Strategy

A successful digital strategy requires more than "dreaming in Technicolor." The ability to develop, implement, and evolve an effective digital strategy is fostered and supported by a number of new organizational capabilities and components, which in turn will be guided by the evolution of an organization's digital strategy. Focus group members and experts both agree that the success of a particular strategy depends less on the technologies involved and more on the ability to implement them innovatively by rethinking strategy, culture, and talent (Kane et al. 2015a; Kiron et al. 2013a). As one researcher explained, "It's not about acquiring technology, but reconfiguring your business to take advantage of the information new technologies enable. Digital technologies must be integrated across people, processes, and functions to achieve an important business advantage" (Press 2016a).

The focus group identified eight capabilities and components which, although not digital strategy per se, need to be in place and interacting with digital strategy in order for it to be a success:

1. *A data and analytics strategy.* Data and analytics are essential to digital strategy for two reasons. First, they support informed decision-making, improved customer experience, and process improvement (Wade 2016; Wixom 2016). Second, the data created by digital

initiatives enable new opportunities that will guide the evolution of a digital strategy. As we have noted elsewhere, there are four major sets of issues related to data that must be addressed: policy, operations, stewardship, and standards. Each of these will guide the decisions that are made about data and must in turn be informed by a specific strategic approach and focus that is set at a high level in the organization. Like digital strategy, a data and analytics strategy is a journey, not a one-off project, which grows and evolves with the direction of the company. A healthy respect for data and the ability to use it well at all levels of the organization has long been shown to contribute to organization performance (Marchand et al. 2000). However, a data and analytics strategy is essential not only to delivering immediate value to the business, but also to putting the pieces in place that will enable new digital business strategies in the future.

2. *Pervasive relationship management.* Digital technologies are putting pressure on companies to provide a unified global customer experience, which in turn helps them engage with the world no matter where they're located (Hirt and Willmott 2014; Press 2016b). Because of this, companies need to rethink their network of business relationships and how they incorporate partnerships to create value. Furthermore, digital business also increases the interconnections between

people, organizations, and devices to enable new products, services, and business models (Blosch and Burton 2016). To take advantage of these new opportunities, companies need to create and reach out to their business ecosystems—organizations, people, and technology platforms—for ideas, information, skills, and delivery assistance.

There are many different ways companies can connect with their broader community, such as (Blosch and Burton 2016):

- *Platform ecosystems* that provide a foundational platform for other ecosystem members to develop complementary products and services.

- *Innovation ecosystems* that access capabilities and talents from outside the organization, often from unexpected fields.

- *Interest ecosystems* that create interest around a company's products and services and that can also serve as a source of new ideas.

- *Commercial ecosystems* that are formed by complementary organizations to deliver products and services.

- *Device ecosystems* that connect consumers and organizations. It is essential that a company understand, build, and manage these ecosystems in order to deliver its desired digital future.

3. ***Supportive culture.*** Digital strategy involves significant business transformation and that means supporting new ways of thinking, working, and leading, said focus group members. Typically, digital strategies require high levels of collaboration across organizational silos and especially between business and IT. Often organizations are risk intolerant and so they shy away from some of the recommended digital implementation activities, such as rapid development and implementation, or experimentation (Kane et al. 2015b). Cultural change is notoriously difficult to achieve in traditional organizational structures and therefore requires significant and focused efforts to incent desired behaviors (Kane et al. 2015a). This starts with leadership, said the focus group. Often the biggest cultural problems arise at the top levels of the organization. Leaders can fail to keep up with new technologies and trends because these lie outside their personal comfort zones. As a result, many employees believe that their leaders don't have the skills and abilities to lead digital change (Kane et al. 2015a) and that few leaders are alert to the threats and opportunities of a changing digital environment (Wade 2016). The focus group stressed that education and clear communication are essential to changing culture. Attention should also be

paid to processes and incentives that inhibit collaboration, experimentation, and risk-taking, as well as data sharing and usage. In short, if cultural change is to be realized, people must understand the new corporate strategy, what the new expectations of their behavior will be, and how changes will be rewarded.

4. *New capabilities.* There is widespread agreement that digital capabilities will increasingly determine which companies create value and which ones lose value (Hirt and Willmott 2014). Companies have begun to recognize that they will need a new set of competencies to develop and deliver on a digital agenda. Most of these competencies require an understanding of both business and technology, and the ability to bridge the gaps between the two that often plague organizations. There is no clear agreement about where these skills should reside, only that they are needed. First and foremost, companies need people who can understand and conceptualize how digital technologies can affect current business models (Kane et al. 2015a). These people need to cultivate "hyperawareness" of their industry, business, and technical environment as well as listen for new ideas from employees, partners, customers, and ecosystems (Kiron et al. 2013a; Wade 2016). Second, organizations need to upgrade their HR practices to identify and acquire the new capabilities they will need both internally and externally. "We are not clear about what

exactly we will be doing in the future," said a manager, "but we can plan to have the skills available to take advantage of them." Many of the skills organizations will need are "soft" skills rather than specific knowledge or technical skills. These include: ability to collaborate and share; willingness to experiment and take risks; ability to work in a fast-paced, distributed environment; and ability to be both flexible and disciplined. Third, organizations must increase efforts to cross-pollinate staff skills in order to expand their awareness of context and broaden their skill set (Press 2016a). For example, IT staff could go on sales calls and interact with customers; and business people could go on vendor visits. Developing these new capabilities will be crucial to the success of any digital strategy, concluded the focus group.

5. *Support for experimentation.* Since there is considerable uncertainty about the potential of digital technology to drive all sorts of business transformation, organizations are being urged to develop practices that would enable them to quickly absorb, test, and adopt emerging technologies (Press 2016a). This is a complex challenge that few companies do well. It involves establishing a mechanism to cultivate, evaluate, and integrate innovation (both business and technical) that addresses three key questions: What is desirable for users? What is viable in the marketplace? and What is possible with technology? Furthermore, it requires that

IT is competent to explore emerging technologies with an eye to asking, What is their future potential? One of the best ways of accomplishing this is through the design, development, and implementation of experiments or prototypes, which are specifically focused on answering one or all of these questions. Experiments require a radical change in organizational philosophy in order to accept failures, learn from mistakes, and quickly pivot, as well as a radical change in IT practices to deliver rapidly and to be flexible.

6. *Flexible architecture.* Architecture is usually the function in IT that is charged with assessing emerging technologies and providing the infrastructure that will support experiments (Smith and Watson 2015). It has traditionally been responsible for ensuring a stable infrastructure and promoting standardization that can enable improved security, privacy, and integration, while reducing costly maintenance and outages. However, in the digital world many business users, tired of waiting for IT to deliver, simply use their credit cards to buy the software and technology they want, thereby circumventing architectural plans. As well, many new technologies are now delivered through the cloud, apps, sensors, and mobile, which have their own infrastructure and standards. Increased flexibility is required for enterprise architects to anticipate and plan for the larger architectural implications of new technologies

(Kiron et al. 2013b) as well as for finding ways to ensure that integration and stability are still a focus (Kane et al. 2015b).

7. ***Rapid development and implementation.*** IT development work is currently undergoing a sea change to accommodate digital technologies, new ways of exploring their value, and new business and customer expectations, said the focus group. This is because the ability to develop and implement new products and services on an iterative basis is essential to the success of any digital strategy (Press 2016a,b). Many IT organizations are now using agile development methodologies, which involve developing deliverable pieces of functionality in short intervals and require active business participation. However, as development productivity improved, a new bottleneck—implementation—has emerged. It now appears that operations functions also need to adopt agile processes to facilitate the rapid introduction of iterative products and services and incorporate experiments. Known as DevOps, this new set of practices incorporates operations staff into a development team to ensure the speedy transition of new development output into implementation. Taken together, agile development and DevOps will result in IT organizations that work more productively and flexibly than the IT organizations of the past.

8. *Improved measurement.* Measurement of the impact of the different components of a digital strategy is a challenge (Kiron et al. 2013b). Nevertheless, it is incumbent on business and technical leaders to begin to design measures that will provide them with positive or negative feedback on their digital initiatives as soon as possible, said the focus group. Members stressed that the metrics involved should be business measures for which both business and IT are held accountable, not technical ones. And with digital initiatives, daily or weekly metrics are most important, rather than the more traditional monthly or quarterly measures (Baggi 2014). The group recommended identifying a small number of metrics and then evaluating and evolving them over time. A key starting point for focusing initial efforts is to ask the key "value questions": What value will be delivered? Where will value be delivered? Who will deliver value? When will value be delivered? How will value be delivered?

Taken together, these new practices and components that are required to successfully execute a digital strategy underscore the scope and nature of the transformation that will be involved with a digital strategy, particularly in IT. Clearly, such significant changes will be difficult to accomplish while at the same time continuing to sustain "business as usual"—something akin to changing the engine of an airplane while in midair, said the focus group. Nevertheless, this is precisely what is being asked of today's organizations.

The challenges are immense, the group agreed. "The field is moving so fast, you can't keep up," said one. "It's very hard to get a shared vision of change," said another, "and piecemeal approaches miss the mark." However, others noted that their organizations have capabilities that they're not taking advantage of. "Our biggest problem is that we need skills to integrate what we already have," said a manager. All participants believed that their organizations must be responsive to the new digital world. "We can't stay where we are," said a manager. "We cannot lead or influence if we are focused on traditional business applications and operations and ways of working."

Conclusion

Organizations have suddenly moved from "IT doesn't matter" to an awareness that digital business will likely disrupt every industry (Lopez 2015; Carr 2003). The radical changes required and the uncertainties involved have created a sense of unease in even well-established businesses. There is no doubt that technology will be at the center of whatever happens in a digital strategy. The question is, where will IT be? In the past, IT functions have been guilty of dragging their heels when radical changes are proposed. Today there is a huge opportunity for IT to demonstrate its value to the organization and to become a true business partner in leading and catalyzing the business. IT leaders and staff must rise to the challenge with vision, education, business awareness, and significant internal change.

References

Baggi, S. "The Revolution will be Digitized." *Journal of Direct, Data, and Digital Marketing Practices* 16, no. 2 (2014): 86–91.

Blosch, M. and B. Burton. "Five Business Ecosystem Strategies Drive Digital Innovation." Gartner Research Report, ID: G00291298, January 12, 2016.

Carr, N. "IT Doesn't Matter." *Harvard Business Review* (May 2003). Digital Strategy Conference Blog. "Defining Digital Strategy: Finding Common Ground." http://www.digitalstrategyconference.com/blog/digitalstrategy/what-is-digital-strategy/, @dstrategycon2015 (accessed July 22, 2016).

Hirt, M. and P. Willmott. "Strategic Principles for Competing in the Digital Age." *McKinsey Quarterly*, May 2014.

Iyengar, P. and L. Mok. "Transform Eight Critical Capabilities to Succeed in a Digital World." Gartner Research Report, ID: G00259971, August 8, 2014.

Kane, G. "How Digital Transformation is Making Health Care Safer, Faster and Cheaper." *Sloan Management Review Digital*, 2015.

Kane, G., D. Palmer, A. N. Phillips, and D. Kiron (a). "Is Your Business Ready for a Digital Future?" *MIT Sloan Management Review* 54, no. 4 (Summer 2015).

Kane, G., D. Palmer, A. N. Phillips, D. Kiron, and N. Buckley (b). "Strategy, Not Technology Drives Digital Transformation." *MIT Sloan Management Review* and *Deloitte University Press*, July 2015.

Kiron, D., D. Palmer, A. N. Phillips, and R. Berkman (a). "Social Business: Shifting Out of First Gear." *MIT Sloan Management Review Research Report 2013*. http://www.sloanreview.mit.edu.

Kiron, D., D. Palmer, A. N. Phillips, and R. Berkman (b). "The Executive's Role in Social Business." *MIT Sloan Management Review* 54, no. 4 (Summer 2013).

Lopez, J., J. Tully, P. Meehan, M. Burkett, and D. Scheibenrif. "Agenda Overview for Digital Business, 2015." Gartner Research Report, ID: G00270684, January 5, 2015.

Marchand, D., W. Kettinger, and J. Rollins. *Information Orientation: The Link to Business Performance*. Toronto: Oxford University Press, 2000.

McGee, K. "Definitional Confusion Slows CIOs' Development of Meaningful Digital Strategies." Gartner Research Report, ID: G00271712, January 29, 2015.

Press, G. (a). "Six Predictions About The Future of Digital Transformation." *Forbes.com*, December 6, 2015.

Press, G. (b). "Six IT Transformation Moves for a Successful Digital Transformation." *Forbes.com*, November 23, 2015.

Shen, S. "How to Develop a Digital Commerce Strategy." Gartner Research Report, ID: G00290525, November 2, 2015.

Smith, H. A., and R. T. Watson. "The Jewel in the Crown: Enterprise Architecture at Chubb." *MIS Quarterly Executive* 14, no. 4 (December 2015).

Wade, M. "The Personalization Paradox." Presentation to the Society for Information Management Advanced Practices Council, May 3–4, 2016.

Willmott, P. "Digital Strategy." Interview. https://www.mckinsey.com/insights/business_technology/digital_strategy, May 2014.

Wixom, B. "Generating Business Value from Data." Presentation to the Society for Information Management Advanced Practices Council, May 3–4, 2016.

Chapter 6

Experimentation

O ne of the newest buzzwords in the world of IT is "experimentation." According to many, the *old* ways of working in IT are not sustainable, given the changing face of business and technology, complete with new competitive forces, new business models, digital strategies, big data, and emerging technologies. These are all converging to pressure companies and, by extension, IT to work faster, smarter, and more effectively (Kane 2016; Blosch et al. 2016). With the growing need for companies to become more digital, while keeping up with—and even anticipating—evolving customer demands, the concept of experimentation has been promoted as a way to test potential responses to market changes without making large risky investments (Schulte and Potter 2014). Correctly done, experimentation drives innovation by picking winners and ruling out less opportune endeavors.

At present, "many companies lack the capabilities for rapid change management and experimentation . . . and this inability will exacerbate dramatically" (Schulte and Potter 2014). IT leaders are therefore struggling to better understand where, how, and whether experimentation can be used in their functions. Culture plays a huge role in this area. "Experimentation is toxic to our conservative organization," said one manager. "IT usually focuses on THE solution so changing is tough," said another. "We've never really defined the difference between innovation and experimentation, so this is a confusing concept," said a third.

These comments highlight several issues related to experimentation and IT, and reveal how this critical stage of innovation can fail if not addressed effectively. First, experimentation is not being introduced in a vacuum. It is part of a whole host of new ways that organizations are transforming to address a number of industry, business, and technology changes, such as mobility, improving the customer experience, the Internet of Things (IoT), big data, cloud services, and new sources of value. Second, experimentation involves risk-taking, which is exactly the opposite of how IT staff have been trained to work. Third, experimentation is not a panacea, but a tool to be used wisely. And fourth, adopting experimentation means dramatically changing how people in IT think and work, and this means adapting many IT processes and, indeed, the whole organization's culture to facilitate it.

This chapter begins by discussing what is meant by experimentation and why an organization would choose to experiment. Next we explore a variety of ways that experimentation fits into the IT organization. Then we look at the experimentation life cycle and how it connects with more traditional IT delivery mechanisms. Finally, we make some recommendations for how IT managers can get started with experimentation in their own organizations.

What Is Experimentation?

Experimentation in business is a strategy to reduce uncertainty and deal with disruption (Berman and Marshall 2014). Or as one focus group manager stated, "It's a tool to explore the art of the possible." There are a variety of definitions of this term. Some are more scientific, such as:

> A process conducted under controlled conditions to ... prove a cause-and-effect relationship [and] support the validity of a hypothesis, theory, principle, supposition, procedure, business case or ... something previously untried with technology." (Schulte and Potter 2014)

Others are more generic:

> Experimentation in a business context is the systematic approach to gaining fundamental insight into the underlying issues of a business opportunity or

challenge. Its goal is to reduce the amount of uncertainty associated with a complex, multi-variable problem. The outcome is insight into what might work . . . and what might not. (Potter et al. 2016)

Although some experiments are straightforward, such as online A/B testing, where one variable in a web site is changed and the response gauged, the vast majority of business problems are much more complex (Thomke and Manzi 2014). This is particularly true when a business is a mix of online and bricks-and-mortar.

Experimentation is also a tool for innovation in IT and is often a prerequisite for introducing a disruptive product, service, or business model (Blosch et al. 2016). As such, it must be used appropriately. There are several reasons for undertaking an experiment, including: testing hypotheses, validating assumptions, and reducing uncertainty. The key is to be clear about what is being tested and what the organization wants to learn and to design experiments accordingly (Thomke and Manzi 2014). Experiments must also be carefully designed to prevent systemic bias, sampling errors, and the Hawthorne effect.[1] These issues highlight the fact that new skills are required for successful experimentation.

Even with these preconditions in place, only about one-third of experiments successfully validate their hypotheses

---

[1] Where participants perform better because they are aware they are being monitored.

(Schrage 2015). The focus group stressed that "success" in an experiment is not always achieving the results expected. "We shouldn't always expect success," said a manager. If an experiment is well designed and carried out, there is much to be learned from failure (Thomke and Manzi 2014). It is therefore important to ensure that failed experiments enter a feedback loop where what was learned leads to new experiments (Schulte and Potter 2014). "We don't always leverage our experiments in this way," said a manager. "An experiment should be just the beginning. The true value comes from analyzing and exploiting the data collected" (Thomke and Manzi 2014). Smaller and more focused experiments can make it easier to learn from and build on the results involved and also to combine results with other forms of customer data (Bingham et al. 2014; Schrage 2015).

In addition to learning, there are other less tangible benefits from experimentation, said the focus group. "Having business and IT work together for the same outcome, being able to acknowledge discomfort, pushing the boundaries little by little, and reframing questions all open organizations up to new ways of thinking and working," said a manager. They also help organizations better understand where value is for their customers and which ideas are the most promising. Experiments also help organizations to "fail fast," which reduces the higher costs and risks of other types of innovation. Finally, experimentation helps organizations relax more about change. "IT always says 'it will take 18 months and cost one million dollars,'" said a member. "With experimentation

I can overcome the natural 'no' from IT and plant seeds of change in our business organization." "I keep saying, 'don't worry, we're just experimenting' to everyone and keep exploring new ideas," said another.

Within the category of experimentation, the focus group identified several different types of experiments:

- *Opportunistic experiments.* These explore individual ideas or responses to competitors' innovations. They are designed to answer the following questions: Can we do it? Should we do it? Can we do it in ways that add more value?

- *Strategic experiments.* These are designed to explore future visions of a business model or new products and services. They can also be focused on testing key assumptions of a new strategy (Sund et al. 2016).

- *Data experiments.* These explore ways of collecting, analyzing, and displaying existing and new forms of data (e.g., from sensors) to provide new insights (Thomke and Manzi 2014).

- *Customer experiments.* These seek to find out what innovations will work best for customers to improve their experience and influence retention or to increase sales (Browning and Rammashesh 2015).

- *Business model experiments.* These explore new modes of delivering an organization's products and services with digital technologies. Typically, they involve digitizing one or more aspects of a company's offering.

- *Intra-company experiments.* These are designed to explore how two or more companies in an ecosystem can work together to deliver value. Typically, they involve companies of different types and sizes, and experiments seek to balance costs and benefits appropriately.

- *Behavior change experiments.* These experiments aim to change internal behaviors by enhancing buy-in to new ideas. "The trick is to let others take the credit for new ideas," said a manager. "For example, we have staff make their own training videos so that they are front and center in a change."

Experimentation and IT

If experimentation is a useful tool for exploring what an organization doesn't know, how does it fit into IT and its work? The focus group identified five areas in IT that should incorporate experimentation into its activities:

1. *Strategy.* Experimentation is often believed to be the antithesis of strategy, as if it were a series of one-off

opportunistic events. But studies show that it is experimentation within strategy that yields the most valuable results (Bingham et al. 2014; Posner 2015). This does not mean that strategy should be inflexible, but that it should provide overall direction and alignment for experiments. Strategy serves as a screen for new ideas and a yardstick with which to measure experimental success (Collis 2016). By being more focused in the opportunities selected for experimentation and more disciplined about which opportunities to approach first, organizations can pursue the most advantageous opportunities in a sequence so that each experiment builds on the learning from the previous ones (Sund et al. 2016; Bingham et al. 2014). Using a strategic vision as a filter ensures that the right questions are asked and empowers local experimentation to refine them over time. Thus experimentation becomes emergent strategy as it identifies current mismatches, gaps, or opportunities to improve an organization's fit in the marketplace (Collis 2016).

The focus group agreed that visionary leadership asking the right questions is essential to effective experimentation. "Leaders can also help us evaluate the results of an experiment and decide whether to end, continue, or amend it," said a manager. "In our most recent experiment, as soon as they saw the results, they were eager to fund more work." The results of carefully structured, sequenced experiments can lead either to radical changes in products or business models or walking away from what initially appeared to be an attractive opportunity (Bingham et al. 2014). "Ideas come from everywhere but strategies are about going somewhere in particular, never about going everywhere." Experimentation within the context of strategy enables a disciplined approach to understanding the nature of opportunities and the linkages among them and helps an organization move in a sequenced fashion from their current state to a desired end state (Hunter et al. 2014).

2. *Architecture.* Architecture has an important role to play in experimentation, said the group. Like strategy, architecture forms a context within which experiments should take place. "We provide a predefined set of tools and open-source software for experimentation," said a manager. "And we still need to govern what happens in experiments." Another added, "Architecture used to just say 'No' in the past when people wanted to do new things. Now we ask why and work with experimenters." Architects can also play a broker/facilitator role for those with good ideas, identifying people who might wish to support experiments and ensuring that all points of view are heard in an experiment. "We set the culture by identifying a go-to person," said a manager. "We also ensure openness to failure and mentor experimenters to ensure they define success carefully and promote their work publicly." Finally, it is often

architects who capture lessons learned so that they can be built upon in the future.

3. ***Development.*** Experimentation should be a precursor to other, more permanent forms of development, so there needs to be a clear understanding of what will constitute success and this will likely not be the same metrics as with other types of development. It is self-evident that experimentation calls for a flexible and iterative approach to development—one that incorporates rapid learning, adjustments, and a cross-functional team (Bingham et al. 2014; Conforto et al. 2016). "Agile and DevOps methodologies have helped us change our culture positively to accept this type of development," said one participant. Yet experiments also need discipline in project management and the design of performance metrics.

Experiments can be threatening to others, both in development and in the business, and communication about their purpose and longer-term benefits to the company is essential as these initiatives take shape (Sund et al. 2016). One manager noted that his company is experimenting with a new digital platform for its products, but there is considerable confusion about its purpose and its development practices. "We know we need a digital platform but we aren't selling much with this experimental one and it is using non-standard tools that won't work in the future, yet there seems to be no plan to change it or turn it off."

This situation underscores two key problems with executing experiments: when to stop them and when to move them into the full development process. By definition, experiments should be short term but the reality is, when business people see something they like, they don't want to turn it off and wait for the 18 month/$1 million solution. "This is our nightmare scenario," said a manager. Yet it is an extremely common one, which emphasizes the importance of developing experiments within an architectural framework and governed by clear success criteria and a formal evaluative process to which all stakeholders agree in advance. Failure to do so means supporting "lame duck" experiments much longer than desirable and eating up resources that could be more effectively used elsewhere (Ballé et al. 2016).

4. *Structure.* Organization structure is not the first thing that comes to mind with experimentation, but it is an important consideration, said the focus group. Many companies are organizationally rigid and find it difficult to fit new experimental units into existing structures, particularly when experimenting with new business models. This is one reason why experiments can fail (Sund et al. 2016). Tension and power struggles for funding and other resources are also common since experimentation typically doesn't fit with existing

prioritization, business case, and resourcing practices. There is no obvious answer to these challenges but they must be dealt with if experimentation is to be successful.

Many companies have consigned experimentation to a lab or an offsite location in the belief that, if freed of organizational constraints, it will flourish (Kane 2016; Applegate et al. 2016). This approach views innovation centers as places where all LOBs (lines of business) can come together to address common problems. According to proponents, the key benefit of this approach is its ability to foster synergies across the business anchored in the belief that innovation is best nurtured away from the mainstream business (McKeen and Smith 2012). Others have created separate but not offsite IT and business units to drive experimentation. Here the goal is to place innovation centers internally, either within IT or within specific LOBs in order to more closely tie IT experiments to "real" problems/opportunities and encourage business buy-in (McKeen and Smith 2012).

Neither solution has been totally successful. Separating out experimentation can solve some immediate resourcing issues and tensions, but in the longer term, organizations still need to contend with the fact that their core business is what sustains them. As John Hegel explains, in most companies there's an unstated agreement that experimental units can do what they want in their sandboxes but if they come back to the core business, they will be crushed (Kane 2016).

Ideally, establishing an organizational structure suitable for experimentation should be an experiment itself. Organizations shouldn't settle too quickly on a structure because this is an aspect of the business that needs to be fully explored and experimented with before learning what works best (Sund et al. 2014). A critical element of any new organization structure is that it supports learning, not just in the experimental group but also in the broader organization. For example, reviews should be used as cross-functional learning events to make sure various stakeholder groups are on the same page (Ballé et al. 2016). After all, the primary goal of experimentation is not to undertake a successful experiment but to support business transformation, challenge entrenched assumptions, and evolve dynamically (Ballé et al. 2016; Kane 2016; Schrage 2016).

5. *Capabilities.* Experimentation is an art, not a science, said the focus group. While organizations must facilitate it, much of the success of experimentation devolves to the people involved. New capabilities are needed in both IT and business, and these will be different from those traditionally needed in the core business (Sund et al. 2016). From a management perspective, senior leaders need to have visionary skills to set the right context for experimentation and process design skills to create a work environment that supports and accelerates learning (Ballé et al. 2016). Leaders must signal

clearly that failure is acceptable and expected since one of the biggest issues holding organizations back from exploring is fear of failure and the belief that failure gets punished (Kane 2016).

Within IT, staff need skills for framing problems, working with data, testing prototypes, and collaboration, in addition to agile methods (Ballé et al. 2016; Conforto et al. 2016). They must market their ideas in new and more accessible ways, such as through graphics and stories, said the focus group. And they must also be able to deliver and learn continuously (Ballé et al. 2016). "Many of our existing IT staff won't successfully transition into this new environment," said a manager. "So we will need to bring in staff with new capabilities and insights from outside the firm."

Teams also must be designed differently to balance a much wider selection of skills including data science, statistics, business analytics, financial modeling, industrial design, innovation management, psychology, and social sciences (Blosch et al. 2016). They need broader stakeholder representation and commitment to more frequent and detailed reviews. The goal should be developing teams that know how to learn—teams that can translate disruptive ideas into experimental questions and hypotheses, grapple with real issues, work within standards, and solve problems creatively, all with the idea of improving an organization's products and services and their fit to markets, production, and its industry (Ballé et al. 2016; Blosch et al. 2016).

Although the focus group strongly believed that IT has a mandate to experiment, it recognized that change is tough for their organizations and the types of change involved in making experimentation a reality are hard to articulate and even more difficult to execute. As with experiments themselves, the best way to become an IT organization that fosters experimentation is to experiment with the IT components involved.

The Experimentation Life Cycle

"One of the things we struggle most with is finding a way to formalize experimentation," said a focus group manager. "We need to understand where experimentation fits into our traditional development and production life cycles." Although there is no clear experimentation life cycle, the consensus is that experimentation is part of the innovation process, falling after ideation—where ideas are collected and evaluated—and before the formal development process (even if it's agile) begins (Browning and Ramashesh 2015). The focus group identified several stages of experimentation that are roughly sequential, even though they may iterate several times (see Figure 6.1):

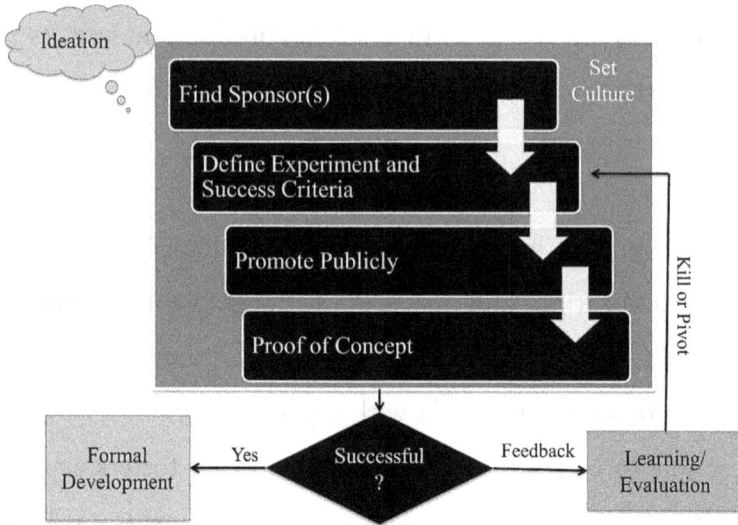

Figure 6.1 The experimentation life cycle

- ***Set the culture.*** As mentioned above, having a culture
  that supports experimentation and its high probability of
  negative outcomes is an essential prerequisite to undertaking experimentation (Browning and Ramashesh
  2015). This needs to come from the organization's senior
  leadership, which will provide the resources, directives,
  and structure for experimentation. Ideally, one business person and one IT person will be responsible for
  experimental initiatives. These people should understand
  the organization's vision for the future, be strong com-
  municators, and have the skills to design and evaluate
  experiments.

- *Find sponsor(s).* Experiments should not be undertaken in a vacuum, said the focus group, but should be consistent with where the company or a business unit wants to move. Finding a sponsor is the acid test of whether an idea has "legs" and will be worth the experimentation. Without one, it is likely that even successful experiments will fail to find traction in the organization. One focus group member found this out the hard way. "We didn't involve the business in our dashboard experiment and when we presented the results to them, they didn't want anything to do with it because they felt threatened and hadn't been involved," she said. Although business sponsorship is the ideal, many participant organizations also had some experiments underway with only IT sponsorship. These were of two types: to test out a new technology or to develop an idea that appears to have promise but which the business doesn't yet "get." Clearly, neither of these can go forward very far without business sponsorship, but occasionally it is worth undertaking a small internal IT experiment so that business leaders can see something that will capture their imagination. "When our leaders saw what we could do with our data, they immediately began asking for more," said another manager.

- *Define experiment and success criteria.* This is a critical step. Experiments need to be focused and have clear success criteria. A design should articulate a hypothesis to test and what result is expected. Early-stage experiments should be simply designed to get very quick feedback, possibly initially from internal users or customer focus groups. Success criteria should also be focused on what the experiment is testing, such as understanding which market demographics would be interested in a new offering. These criteria should be clearly agreed on in advance by all stakeholders as they will form the basis of the learning and evaluation stage.

- *Proof of concept.* The goal of this stage is not to build *the* solution but a minimal viable product to test an idea and see if it makes sense. Iterative development is essential for experiments, although very early stages might not be technology based but could be mock-ups or storyboards. As with agile development, speed is of the essence, but iterations will likely be even more rapid with experiments, since there is much less certainty.

- *Learning and evaluation.* This stage is the focus of the experimentation process for several reasons. First, it is critical to assess the results of the experiment. What happened? What do the data say? Often, it will be important to dig down into the results to determine what can be learned from an experiment. Common pitfalls include: looking only at overall outcomes, not at specific market segments; canceling the experiment too early when uptake is slow and not understanding the underlying reasons for this; and dragging the experiment on too long when initial results are not borne out over a longer time period. A second reason to carefully assess results is to determine if there were design or planning issues that undermined them. Often human error and systemic bias can lead to less-than-clear results. Such situations can feed back into the experimental design stage.

Third, what is learned will help direct future experiments. "Pivoting" is a key component of experiments as leaders increasingly understand more clearly how to develop an idea. Some key types of pivots include: zooming in on a particular feature; refocusing on a particular customer segment; reframing the business model proposition; changing the mix of channels involved; and changing the technology involved (Blosch et al. 2016).

Fourth, experimental results should be assessed with regard to the overall process of strategic transformation. Success could lead to the development of new products or services but all results will feed back into the strategy development process and dynamically help guide business strategy and transformation. Success could also lead to a series of related experiments to test improvements to existing products and services and/or features or new products and services. The goal is to conduct an ongoing number of experiments within a particular strategic context, rolling out the successful ones and re-evaluating the not-so-successful ones. Finally, learning should lead to an assessment of the skills, capabilities, and resources needed. Is more business participation needed? Is funding adequate? What new capabilities should be added to the team? Are there cultural or procedural roadblocks? Each of these questions should be addressed as part of this stage and the results fed back to sponsors and the leaders responsible for experimentation.

Driving Discovery with Experimentation

As with all new ways of working, experimentation will take time to become engrained in an organization's culture. The focus group had several recommendations for IT leaders who wish to get this change started:

- *Provide space and time for experimentation.* Experimental thinking uses a different type of brain power that requires a disengagement from current tasks and activities (Posner 2015). Nurturing it requires people to gain a psychological distance from more performance-oriented work. Companies that wish to promote a culture of experimentation need to provide places and spaces for it—both physical and virtual—where people can take time away from their day-to-day activities to think, interact, cross-pollinate ideas, and design experiments related to different topics (Berman and Marshall 2014). This does not require a separate location and dedicated resources, at least initially, but it does mean recognizing and respecting that a certain percentage of everyone's time should be devoted to this type of work. One focus group organization has dedicated Friday afternoon as "thinking time" to be used for learning. "I don't care what my people do with it as long as they bring back something they've learned," said a manager.

• *Use cross-functional teams.* "I've learned the hard way that you have to involve business in experimentation or whatever you come up with is likely to fail," said a manager. The best way to do this, said the focus group, is to start a relationship with colleagues in the business. "Use food and drink," said a manager. "Spend some time cultivating them. Be humble and ask for their ideas and opinions. Be a little fuzzy if you've got an experiment underway so they can help." Many focus group companies now participate in offsite innovation labs, which involve a combination of technology and business partners. Others prefer to create internal cross-functional experimental teams that can seed changes back to the broader organizational culture, helping it learn from failure and become more comfortable with uncertainty (Blosch et al. 2016). Some experiments will also involve customer participants through focus groups.

• *Establish new ways to fund and govern experiments.* It is clear that experimentation cannot and should not compete with other types of IT projects, and that organizations need to develop quick and effective ways to engage in, fund, and disengage from experiments (Blosch et al. 2016). One focus group company has created a combined innovation council. "We've recognized that we can't do things in silos so we've brought the business and IT together to experiment with how to manage experiments more effectively," said a manager. Another organization uses a "dragon's den" approach to experiments, where their proponents pitch their ideas to business leaders to obtain preliminary funding. "This emphasizes that experiments must be tied to a business problem," said the manager involved. All agreed that investment should be tied to individual iterations. "That way, we always ensure that we are going in the right direction for our company," explained a focus group member.

- **Reduce known unknowns.** Experiments actually comprise two sets of unknowns: things we don't know we don't know (unknown unknowns) and things we know we don't know (known unknowns). The key to an effective experiment is to reduce the knowable unknowns by focusing attention on things that can be made more certain, such as results, processes, communication, goals, vision, and requirements (Kane et al. 2015). "Many of our experiments fail because of human factors, such as poor planning, rushing design, and other things," said a manager. The remaining known unknowns are driven by complexity. Reduced complexity can be achieved through experiment decomposition and thorough planning, as well as scenario analysis and frequent communication. Experts also recommend using long interviews with users to pick up weak signals, using data mining to better understand the phenomenon to be studied, and incorporating a balance of local autonomy and central control into an experiment's design (Kane et al. 2015).

• **Rethink the role of failure in the enterprise.** Celebrating failure and what has been learned is central to developing a culture of experimentation, agreed the focus group, but it can be hard to sell this concept to the organization. Many leaders consider an experiment a failure if it does not bear out its hypothesis. Changing the culture means getting comfortable with failure and accepting some risks. Often the pressure to claim success keeps many experiments going on too long (Blosch et al. 2016). "The worst thing is when you can't wind down an experiment and there are organizational politics associated with it," said a manager. Senior leaders therefore need to openly talk about failures and then deploy what has been learned in future experiments. This is the best way to keep more ideas and innovations flowing, agreed the focus group.

• **Build on what you learn.** "We had an experiment that wasn't a business success but when we looked at the results we realized we could use the technology involved as a platform for other products and services," said a manager. "That learning was really key and, of course, we learned that this particular business opportunity should not be pursued as well." Focus group managers recommended experimenting first with IT staff as customers and then with internal staff as it gives them firsthand experience before taking a bigger risk with customers. "We experimented with mobile business intelligence tools with carefully selected participants," said another. "But others caught wind of it and it became a movement. Then the Board heard of it and wanted it too so we were able to experiment further. It was perfect!" Although not all experiments are successful in proving their hypotheses, they should all inform strategy. The key is learning from the results and then pivoting. The focus group stressed that experiments should be seen as a journey, not a road map. "Miracles come in small steps," said a manager. "You learn and then you adjust."

Conclusion

Few individuals or organizations are comfortable with experimentation but most recognize that it is necessary for survival in the brave new world of the future, when we will be bombarded with increasing amounts of change from all fronts. There are lots of ideas in organizations, but until recently there has been little support for following up on them and few ways to do so. In this chapter we have shown that introducing experimentation into an organization must be done holistically. Culture, capabilities, strategy, processes, governance, funding, HR practices, and functions must all be on board to do it successfully. Learning from failures is essential. As a result, experimentation is not for the faint of heart. But the fact remains that organizations *are* experimenting and obtaining results, and this increases pressure on others to do the same. Developing a culture where experimentation is accepted and even expected is the first step. Learning to experiment effectively will take longer and may require some process and structural experiments as well. However, the payoff will be the ability to navigate much more confidently in a world of continuous and dynamic change.

References

Applegate, L., and A. Meyer. "Launching 1871: An Entrepreneurial Ecosystem Hub." *Harvard Business School Case N1-816-090*, April 20, 2016.

Balle, M., J. Morgan, and D. Sobek. "Why Learning is so Central to Sustained Innovation." *MIT Sloan Management Review* 57, no. 3 (Spring 2016).

Berman, S., and A. Marshall. "Reinventing the Rules of Engagement: Three Strategies for Winning the Information Technology Race." *Strategy and Leadership* 42, no. 4 (2014).

Bingham, C., N. Furr, and K. Eisenhardt. "The Opportunity Paradox." *MIT Sloan Management Review* 56, no. 1 (Fall 2014).

Binns, A., B. Harreld, C. O'Reilly, and M. Tushman. "The Art of Strategic Renewal." *MIT Sloan Management Review* 55, no. 2 (Winter 2014).

Blosch, M., N. Osmond, and D. Norton. "Enterprise Architects Combine Design Thinking, Lean Startup and Agile to Drive Digital Innovation." Gartner Research Report, ID: G00295415, February 4, 2016.

Browning, T., and R. Ramashesh. "Reducing Unwelcome Surprises in Project Management." *MIT Sloan Management Review* 56, no. 3 (Spring 2015).

Collis, D. "Lean Strategy." *Harvard Business Review*, March 2016.

Conforto, E., E. Rebentisch, and D. Amaral. "Learning the Art of Business Improvisation." *MIT Sloan Management Review* 57, no. 3 (Spring 2016).

Hunter, R., M. Mesaglio, and O. Chen. "Experimentation is Key to Strategy in the Era of Digital Business." Gartner Research Report, ID: G00260164, September 17, 2014.

Kane, G. "The Dark Side of the Digital Revolution." *MIT Sloan Management Review Digital*, 2016.

Kane, G., D. Palmer, A. Phillips, and D. Kiron. "Is Your Business Ready for a Digital Future?" *MIT Sloan Management Review* 56, no. 4 (Summer 2015).

McKeen, J. D., and H. A. Smith. *IT Strategy: Issues and Practices (2nd ed.).* Upper Saddle River, NJ: Pearson Education, 2012.

McKeen, J. D., and H. A. Smith. *IT Strategy: Issues and Practices (3rd ed.).* Upper Saddle River, NJ: Pearson Education, 2015.

Posner, B. "Why You Decide the Way you Do." *MIT Sloan Management Review* 56, no. 2 (Winter 2015).

Potter, K., M. Smith, and S. Buchanan. "Definitions Required to Guide Business Unit IT Strategies." Gartner Research Report, ID: G00299635, April 21, 2016.

Schrage, M. "Embrace your Ignorance." *MIT Sloan Management Review* 56, no. 2 (Winter 2015).

Schulte, A., and K. Potter. "CIOs Must Drive Rapid Change and Experimentation in the Enterprise for Greater Innovation and Competitive Advantage." Gartner Research Report, ID: G00260204, November 5, 2014.

Smith, H. A., and J. D. McKeen. "Developing a Digital Strategy." http://www.itmgmtforum.ca, 2015.

Sund, K., M. Bogers, J. Villarroel, and N. Foss. "Managing Tensions Between New and Existing Business Models." *MIT Sloan Management Review*, Summer 2016.

Thomke, S., and J. Manzi. "The Discipline of Business Experimentation." *Harvard Business Review*, December 2014.

Section III

Delivery Drivers

Innovation must be delivered rapidly and continuously, while still ensuring high quality deliverables, monitoring value delivered, and complying with security, privacy, and governmental regulations. In Chapter 7, "Redefining IT Governance," we discuss the new IT governance practices that are needed to facilitate nontraditional approaches to technology delivery. Chapter 8, "Managing IT Disruption," examines the many ways that IT organizations are themselves being disrupted while also being expected to deliver disruption to the business as whole. Finally, Chapter 9, "DevOps," describes how many different IT functions and processes must be restructured to cope with new ways of implementing technology rapidly and successfully.

Chapter 7

Redefining IT Governance

IT is facing many new pressures to deliver faster, deliver different products and services, and work differently at all levels of the organization and with vendors, partners, and even customers. As a result, there is broad recognition that IT needs to redefine itself to be more effective for the future. Change is not optional for today's IT organizations. The major challenge for IT leaders is therefore deciding what to change and how fast to do it.

The pressure is on IT to deliver more quickly and efficiently, and this means working differently. IT managers are in general agreement with this goal but also feel a responsibility to protect the organization and its data. Many of IT's so-called bureaucratic processes were put in place for good reasons, such as to ensure quality, interoperability, and cost-efficacy. And in many highly-regulated industries, such as finance and healthcare, laws and risk aversion govern much of what can and cannot be done by IT. Nevertheless, there is a need to reconcile these competing priorities and rethink how IT works if the goal is to drive innovation.

IT work involves two major components: 1) making decisions about what work to do (i.e., strategy), and 2) delivering the work (i.e., execution). IT governance is the system of structures, processes, and roles that collectively *oversee* these two major components of IT work. Championing the needs of the enterprise (i.e., common processes, architecture, data, and controls) favors more centralized forms of governance just as championing responsiveness to the business favors more decentralized (or perhaps hybrid) forms of governance. Balancing these is at the nub of the challenge for today's IT leaders: the need to act faster and in closer alignment with the business while still protecting the organization's overarching interests.

The Increasing Importance of Governance

IT governance is a framework of processes and structures that specify who makes decisions about and who is accountable for the IT function and its work. It also determines who should have input to issues, how disputes should be settled, and how decisions should be made, implemented, and managed (Weill and Ross 2005). It is *not* about what specific decisions are made or how groups are organized and led. Effective IT governance is designed to encourage desirable behavior in the use of IT that is consistent with an organization's mission, strategy, and culture (Weill 2004).

Research shows that effective governance has a significant influence on the benefits an organization receives from its IT investments. Value is achieved by ensuring "that the right groups are making the key IT decisions so that those decisions enable the desired goals and behaviors of the enterprise" (Weill 2004). Although it does not point to a single best governance model, effective governance is carefully designed to link to an organization's particular performance goals (Weill and Ross 2005).

A significant reason for an organization's ability to derive value from IT is that its governance provides senior leaders with a clear understanding of how IT decisions are made, thus helping to: 1) clarify business strategies and IT's role in achieving them, 2) measure and manage IT investments, 3) design organizational practices to align IT and business strategies, 4) assign accountability for change, and 5) learn (Weill 2004). Therefore, it is unfortunate that IT governance has all too often been found to be a mystery to key decision-makers at most companies (Weill and Ross 2005).

Although getting value from IT is an important reason for leaders to design for effective governance, in recent years a number of other factors have also become drivers for senior managers to focus on it. These include:

- *Ensuring privacy and security.* Concerns about security have now reached the board level. More and more, CIOs are being asked to present their IT security plans to directors, who recognize that a security breach could at minimum embarrass their company, and potentially cause significant losses. Recent accounts of major customer information losses due to security lapses have heightened attention to the need to have effective governance of security practices even in the most insignificant areas.

- *Compliance with laws and regulations.* There was general agreement in the focus group that there is a growing amount of legislation affecting governance, that regulators in specific industries are becoming more demanding, and that both internal and external auditors are becoming more intrusive and prescriptive in their reports. "This all results in more process and additional work," said a manager. While not all industries are regulated, many are. In this focus group, companies from the finance, insurance, healthcare, travel, and food industries all noted the increasingly onerous burden of regulation. And all companies are feeling the pressure of new legislation and more detailed audits. In addition, international or global companies must comply with a variety of individual country regulations, such as separating data or management oversight.

- *Improving risk management.* IT work has become more complex and is less often under direct management control. Today's IT service offerings typically include third-party software developed by outsourced staff (often not even in the same country) and rely on a rapidly evolving ecosystem of service providers in emerging industries (e.g., cloud, software-as-a-service). Under these circumstances, it is all the more important for governance to recognize and address the additional vulnerabilities involved.

- *Improving alignment between strategy and execution.* Often organizations have misaligned governance structures—one for innovation and strategic projects and another for execution and operations. If actions in one ignore governance requirements in the other, such as when putting new changes into operational systems or circumventing existing architecture, then governance is undermined to the detriment of the company as a whole.

- *Increasing customer involvement.* As technology touches the lives of end customers more often and is more visible, corporate reputations are increasingly at risk. As one manager put it: "It's important that we use customer information to interact with our customers appropriately or it could be embarrassing or worse."

Group members commented that with so many competing dynamics, it is especially difficult to design governance without adding significant extra work for IT staff. "Our goal is to design governance that is *enabling*," said a member. "If everyone understands their roles and responsibilities, when they need permission, and the right people have the right tools, then this is possible. Not everyone will like it but at least it is clear." Another added, "Our goal is to make effective governance a part of our culture. We want to build integrity into everything we do."

It is clear that governance is more of a challenge the larger a company gets, and it's a greater challenge in some industries or countries than in others. Regardless of the company, the need for effective IT governance has never been more evident, and companies are scrambling to keep up as IT itself changes and the interaction between these drivers evolves.

Elements of Effective IT Governance

Effective governance is not attained by a single committee or set of rules, except perhaps in very small organizations. Instead, it is achieved through an integrated framework of organizational groups that provide oversight in different areas, standards, and practices founded on industry best practices, legal and compliance requirements, and policy guidelines that direct how work is to be accomplished. Governance should be designed to provide clarity and consistency to IT work and focus IT decisions on what is most important to the organization. It should also ensure that governance practices support each other, rather than work at cross-purposes.

A governance framework most often operates at three levels, although this may vary according to size and geographic needs:

- *Board governance.* While board level governance involves more than IT, boards are increasingly aware of their responsibility to become better informed about the IT decisions made in their organizations. Boards are responsible to their shareholders for managing both the finances of their companies and the risks they undertake, as well as their reputation and brands. In addition, a board is responsible for ensuring that its organization is compliant with all regulatory requirements and reporting, and for addressing all issues raised by its internal and external auditors. Each of these can now be significantly affected by IT strategy and execution. One need only pick up a newspaper these days to see how damaging it is for a board to fail in its oversight of one or more of these areas. Thus many boards now have technology committees or at least members who have some technology background, and IT matters are being more frequently questioned at this level. This is a positive change according to the focus group, as it raises the profile of serious IT concerns to the highest echelons in the firm.

- *Enterprise IT governance.* This is a level of governance that is growing considerably as IT leaders realize the value that good governance can deliver. As Figure 7.1 shows, enterprise IT governance provides detailed integration of factors affecting IT decision-making in several areas including:

  - Architecture. This sets current and future technology strategy, ensuring it is consistent with overall business strategy. It identifies best practices and industry standards for use in IT and provides oversight and approvals for all technology initiatives to ensure that the company is both protected and positioned well for the future.

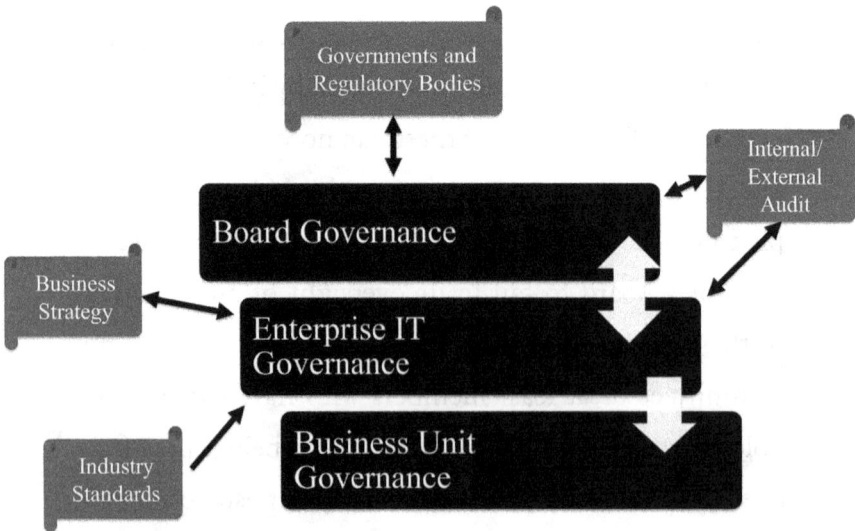

Figure 7.1 IT governance integrates many components.

- Security and privacy. This provides oversight and guidance on all security matters to ensure that the organization's information assets and infrastructure are protected and to proactively and continuously address security risks. It also oversees access management practices.

- Operations. This ensures that all technology implementations and changes follow proper procedures and standards, and provides oversight of all ongoing operations. It also reviews and reports on ongoing service levels and seeks to ensure that risks are mitigated and issues resolved.

- Strategic projects. This oversees the processes designed to ensure that the most strategically valuable projects are properly and expeditiously resourced and funded. It also tracks project progress to ensure that all standards are followed and risks identified and mitigated.

- IT capabilities. This ensures that IT has the necessary capabilities to carry out its mandate, whether internal or external, and also oversees the organization's relationship with its vendors and service providers, ensuring they meet all standards and contractual responsibilities.

- Data. This is the newest component of IT governance, responsible for managing data as a strategic asset, building a common framework and definitions for key corporate data, and developing effective data management practices and accountabilities.

• *Local (business unit) governance.* Most organizational governance frameworks also allow for some discretionary IT decision-making. This involves making decisions about smaller-scale, lower-cost, new applications and prioritizing changes to existing IT applications to be made in the business units. In general, organizations try to limit business unit IT decisions in order to focus the vast majority of IT spending on higher-value, strategic initiatives. And most recognize the need for this level of decision-making to support local flexibility. Enterprise IT provides oversight for and input to this level of governance, ensuring it follows appropriate practices and standards for development, testing, and implementation.

IT Governance Evolution

Once designed, IT governance should not need to be rethought simply because of changes in the economy or adjustments to strategy (Weill 2004). Circumstances under which governance should be completely re-evaluated include a major change in strategy or a merger. Within these two extremes, however, lies the area of "practical governance," which involves enhancing, extending, or tweaking governance models to provide more clarity or to adapt to changing IT practices. "We have evolved our current governance model over the past six or seven years," said one manager, "and we still have areas that we need to address more fully."

Evolving areas of governance usually lie in new areas of IT work where best practices and standards are not well understood, such as:

- *Third party vendors/outsourcers.* Many companies have found gaps in governance and assumptions about decision-making when work is subcontracted out to third parties, who may in turn subcontract out some work. This can lead to any number of problems from a legal, audit, compliance, security, or privacy perspective. For example, "We thought we could outsource application development to get it done more quickly, but then we found we had to retrofit these applications to meet our security criteria. This was very painful," said a manager. "Because IT still has accountability for products developed in this way, we've found it necessary to add extra layers of due diligence when dealing with third parties," said another. Although it introduces additional complexity, there was general agreement in the group that with a good governance framework in place, and if vendors understand it well, shifting IT work to third parties can be successful.

- *Cloud services providers.* "With cloud software-as-a-service, we at first naively assumed we could impose our own controls, but with these large cloud vendors, *you* have to adapt," said one manager. Under these circumstances, companies must be extra vigilant to ensure that cloud vendors meet all corporate, regulatory, and government criteria, such as where data is located. Many countries explicitly prohibit their data from virtually or physically leaving their borders. In these cases, companies must evaluate their own processes and policies as well as those of their vendors to both adjust their internal governance practices and ensure that certain minimum standards are adhered to by their provider. "If these can't be met, you shouldn't be doing business with them," said a manager.

- *Mobility and other new technologies.* Although new technologies do not affect governance per se, there are at least three reasons why governance needs to be reconsidered when they are introduced. First, the technologies themselves may contain new vulnerabilities that have never been considered. For example, company tablets may contain confidential information or customer data that could be stolen or hacked due to immature protections on these devices. Second, in the rush to implement new technology and innovative applications, a company may be pressured to short circuit tried-and-true practices that could cause anything from operational issues to simple malfunctions to occur. Many companies are experimenting with new technologies these days and so developing ways to explore potential opportunities safely without circumventing existing governance practices is a challenge for IT leaders. One company in the focus group has a "fast track" approval process that enables strategically desirable projects to jump the queue in the prioritization process, while still following the standard review procedures afterward; another has developed a "governance-lite" approach for pilots, with the understanding that if the decision is made to scale these up into production, they will follow all proper procedures. Third, new technologies typically have immature and rapidly changing standards that make it difficult to integrate into a technical architecture. An IT nightmare can occur if different projects select similar new technology that operates in different ways. Good governance prevents this from occurring.

- **Data.** The most challenging area of governance at present for many companies is data, and it is often a political hot potato. "Data has been a very painful journey," sighed one manager. "We've tried to do it many times before. We're now confident we have the commitment and the roles, responsibilities, accountabilities, and processes clear, but we are still meeting bi-weekly to figure out what's not working and who's not engaged so that we can deal with it. Otherwise, we'd be wasting our time." Good data governance requires consistent standards for each piece of data, but without clarity about who owns what, who produces it and who uses it, the information produced can be inaccurate or conflicting. One company learned a hard lesson when its CFO asked for integrated revenue data only to learn that different business units had different mechanisms for calculating this figure, which made the information highly suspect. "We needed to start top down and determine who owned which data and who had a mandate for changing it," said the manager involved. "We had a good data architecture for sourcing data but needed governance. We are now looking to develop data standards and good management practices."

In these new areas, governance best practices are not yet mature, and companies are trying to adapt their existing governance models to fit the need. Managers in the focus group stressed the importance of monitoring new governance practices as they evolve to ensure they are working well and doing what they are supposed to do. Without this attention, governance models can fall into disuse or become so bureaucratic that a company is seriously impeded in its work. "Good governance should help everyone understand the rules and make sure that the right levels have the right tools and responsibilities. We should never lose sight of this when designing governance," said a manager.

Driving Delivery with Redefined IT Governance

Focus group members had several recommendations for managers seeking to re-design or improve their IT governance practices:

1. *Make fact-based decisions.* "Ideally, all decisions should be based on facts, not gut feel," said a manager. "But we still have some way to go in this area." The group agreed that the more facts that are brought to bear on a decision, the better that decision will be. Facts improve clarity and consistency and create trust in the decisions that are made. As a result, they also encourage people to play by the rules and reduce "politicking."

2. *Work from the premise that one size does not fit all.* Governance in any company should be as simple as possible with a limited number of goals (Weill 2004). However, the group emphasized that governance will vary according to the industry, size of company, and geographical makeup of the organization. Thus a global financial company will need a more complex governance model than a small firm in a less-regulated industry. Many companies have been successful with different governance models (Wade and Buttchel 2013). "Small companies and large ones have the same set of issues, but they need to deal with them differently," said one

manager. Similarly, centralization is not necessarily better than decentralization. "We should aim for the right model for the right types of decisions," said a manager. "We shouldn't centralize simply for the sake of centralizing," added another.

3. *Monitor and iterate.* Although governance models should be carefully designed to reinforce an organization's goals, they are not easy to get correct the first time. The group stressed the importance of monitoring how a model works in practice and making adjustments to ensure all processes integrate smoothly and efficiently. "It is especially important to align strategy and execution governance," said a manager. There can also be resistance to governance that is perceived to be inhibiting. The group stressed understanding where governance is not working and why, adapting it in some cases and reinforcing the rules in others. "At some point, you have to say 'these are the rules' but you also need to make sure that the culture is aligned top-down to meet these goals or governance will be perceived as intolerably bureaucratic," a manager commented.

4. *Communicate from the top.* It is essential that the CIO and senior leadership team communicate strategy clearly throughout the organization and educate all levels of personnel about the rationale for key governance mechanisms. Ideally, governance should be able to be

communicated on a single page and education should be used to better align the culture with organizational objectives. "Often we see that our VPs are aligned with our goals but this doesn't filter down to other levels where our projects are actually implemented. We therefore need a much heavier focus on communication and education to reinforce governance," said a manager.

5. *Clarify strategies and principles for staff augmentation.* With so much IT work now being done by vendors, it is critically important to ensure that an organization has clear strategies and principles for how staff augmentation is to be handled. Organizations need to ensure that vendors comply with their governance practices and that staff is fully trained in what is expected. According to one manager: "This is especially important when dealing with offshore companies who may have different assumptions from our own. We've learned that we have to select vendors very carefully and hold them to very high standards. In our company, we rigorously review all outsourced work to ensure staff have followed our practices."

6. *Include release valves.* Even the best designed governance model needs an exception process to handle justifiable deviations from best practices, said the group. "Standards are black and white," said a manager. "We need to have ways to exempt projects from them

if there's a good reason, such as a compelling business opportunity." However, exemptions should only be made after a clear consideration of the potential impact and risks, and often for a limited time period. This might be the case if a project wants to use a new and untried technology or if a competitor has come out with a new product or service that has rapidly become "table stakes."

Conclusion

Effective IT governance is an essential element of delivering IT value. Designed well, it can facilitate alignment with corporate strategy and performance objectives and enable best practices in risk management, security and privacy protection, auditability, and information management. It ensures that the right decisions are made by the right people at the right time and provides guidelines for how best to address redefining IT. Designed poorly, it can be a roadblock to innovation, hinder performance, encourage non-compliance with the rules, and wrap the organization up in red tape. CIOs play a very clear role in setting the right tone at the top, promoting education about the role of governance, and creating transparent processes based on facts. There are many excellent governance guidelines available for IT leaders to use as a starting point, but they must also make the time to ensure that their governance actually works in practice and make adjustments where needed.

References

Wade, M., and B. Buttchel. "Anchored Agility: How to Effectively Manage the Balance Between Local Flexibility and Global Efficiency." *SIM Advanced Practices Council Research Report*, January 22, 2013, http://www.simnet.org.

Weill, P. "Don't Just Lead, Govern: How Top-performing Firms Govern IT." *MIS Quarterly Executive* 3, no. 1 (March 2004).

Weill, P., and J. Ross. "A Matrixed Approach to Designing IT Governance." *MIT Sloan Management Review* 46, no. 2 (Winter 2004).

Chapter 8

Managing IT Disruption

"**D**isruption" is undoubtedly a hot topic. It describes a number of different but related challenges facing both business and IT leaders. First, industries and companies are facing new threats from nontraditional competitors using new technologies in new ways to scoop up their most profitable revenue streams. As a result, "disruption has moved from an infrequent inconvenience to a consistent stream of change that is redefining markets and entire industries" (Plummer et al. 2016b). One survey of board of directors members found that 32 percent believe their company's revenue will be threatened in the next five years and 60 percent want to spend more to directly address these potential disruptions (Weill and Woerner 2015). And that means more work for IT, which must play catch-up.

Second, new technologies are being continually developed and brought to market. Business leaders know that in order to stay relevant, they must constantly survey these emerging tools and techniques to determine which they should adopt to deliver new forms of value to stay ahead of the curve (Weldon and Rowsell-Jones 2015). And this means that IT is expected to participate or even lead these evaluations and analyses, and develop proofs of concept.

Third, with all these new technologies and new applications comes the threat of new disruptions—either malicious or accidental—to an organization's existing infrastructure and legacy environment (Sheffi 2015). Since IT is also charged with ensuring business continuity and information security, this has become no small task now that online devices are everywhere.

Finally, IT is faced with disruption in its own delivery models from cloud computing, global outsourcing and managed services, the need for continually changing new skills, and ongoing pressure from business to deliver faster, better, and cheaper. With disruption coming from every direction, it is paramount that IT leaders address this situation in order to deliver innovation successfully.

"The good news is that because of all this disruption, our organizations now understand the new value of IT," said a senior IT manager. "They have much bigger appetites for new technology than a few years ago." "But the bad news is that disruption can be a noun, a verb, or a profanity depending on the situation and what you are talking about," explained another. The focus group noted that although disruption is challenging many fundamental assumptions, it is also an opportunity that can be managed. "The key is ensuring that we are proactive wherever possible, not reactive," said a third manager. "Change is a choice that can be planned; disruption is not a choice and is therefore chaos."

IT organizations are stepping up to these challenges in a number of ways, including implementing new methodologies such as agile and DevOps, developing new capabilities such as thought leadership and analytics, and examining new ways of understanding the value of IT, such as through experimentation. However, there is still a sense that this is not enough and that IT is struggling to keep up. In this chapter we explore the concept of disruption in IT and how to effectively deal with it. In the first section we look at the nature and sources of disruption in organizations and how these affect IT work. Next we look at how IT is responding to these and to disruptions in IT work itself and evolving to become a new type of technology organization. Then we examine what the IT of the not-so-distant future might look like as a result of today's disruptive forces. In the final two sections we discuss best practices in managing disruption and some steps managers might take to help turn the chaos of disruption into the opportunity of change.

Disruption in Organizations

The term "disruption" usually means a change or disturbance to a current or settled way of doing something (Yokelson et al. 2016). Although technology is not always *the* cause of disruption, it is generally involved in some way. Often disruption is the result of a combination of the new capabilities offered by a technology, complementary products and services that build on that technology, and changes in the expectations of and behaviors in a culture, market, or process as a result (Yokelson et al. 2016). The telephone, for example, was a new technology that was not disruptive of current ways of working or living until it was made widely available to businesses and members of the public and until phone companies provided the network of wires, operators, and service people that made it possible for people to connect with each other in an entirely new way. It may be many years before the full impact of a new technology is apparent, as new ways to incorporate it or to adapt behaviors and structures evolve. These secondary ripple effects can be even more disruptive (Plummer et al. 2016a). Thus with the addition of mobility and online connectivity the telephone is still a disruptive influence, albeit one that was never anticipated by its inventors or early users.

Today disruption is a more serious concern for organizations than it has been in the past because of the vast number of possible changes enabled by a literal explosion of new technologies and new forms of connectivity. These changes are so rapid that it is almost impossible to keep up with them, let alone determine which ones might be useful or lead to fundamental alterations in the processes of creating, producing, and delivering an organization's products and services (Plummer et al. 2016a). Current business environments are fraught with hype and noise about disruption and new technologies, making it extremely difficult for organizations to see the truly disruptive threats and opportunities (Smith et al. 2016). Because of this, organizations have become fearful that they are vulnerable to competition from unexpected sources (Gans 2016) or risks from unanticipated events (Sheffi 2015). Many are aware of Christensen's classic study of disruptive innovation, which found that successful businesses can fail in the face of technological change because they tend to ignore innovations in favor of following the paths that have made them successful. Essentially, they don't move forward until it is too late (Gans 2016). And all leaders are aware of companies that have not anticipated some of the risks and vulnerabilities associated with their technology and whose businesses have subsequently been damaged.

In short, companies are concerned that new technologies or new uses of existing technologies can be utilized by a non-traditional competitor to develop innovative capabilities, products, or services that will undermine their traditional markets and sources of value, thereby making their current business models and assumptions obsolete (Plummer et al. 2016a; Berman and Marshall 2014). These disruptive forces create new connections, linking institutions and individuals in new ways, breaking down established boundaries, and making it possible to orchestrate complex value chains through ecosystems of organizations (Berman and Marshall 2014; Grossman 2016). Often, it is difficult for an existing organization to catch up because of its embedded culture, processes, and systems that are designed for its current structure and activities.[1] Figure 8.1 summarizes these new disruptive sources.

Because technology is at the root of most disruption, it is natural for business leaders to look to IT for guidance, insights, leadership, and protection. Ideally, the focus group agreed, it is IT's role to bring new technology opportunities to the business, help its leaders distinguish between hype and true disruption, and protect the organization from information security risks. "Our business is increasingly exposed to disruption from new technology," said a manager. "Our customers' increased appetite for digital capabilities is causing a lot of disruption for us," added another. Unfortunately, "our company faces hundreds of industry pushes and pulls," said a third. "We in IT simply cannot make these changes overnight."

Speed of response is central to dealing with disruption and can spell the difference between planned change and survival (Sheffi 2015). The focus group and others agree that disruption can be managed if it can be anticipated and responded to quickly and efficiently (Gans 2016). In fact, disruptors themselves may not be in the best position to take advantage of their ideas. If established organizations can recognize the

---

[1] "Conway's Law" holds that any organization that designs a system will produce a design whose structure is a copy of the organization's current communication structure.

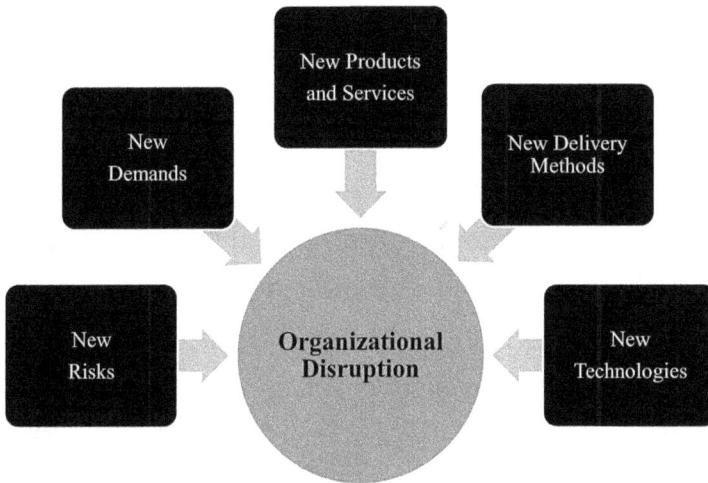

Figure 8.1 Sources of organizational disruption

indicators of a disruption early, they can bring their resources and complementary assets to bear and shift advantage to themselves (Plummer et al. 2016a). How to do this effectively is the real challenge for organizations.

Disruption and IT

IT is called upon to assess new technologies, develop new products and services in response to customer and business demands, support new delivery channels, and manage the risks involved. Yet while it is supporting the business in dealing with disruptive forces, IT is also in the midst of being disrupted itself. The same types of forces affecting the business also affect IT functions independently of the business.

Many operations groups are being disintermediated by cheaper cloud-based infrastructure-as-a-service or outsourcing arrangements. Similarly, development functions are being displaced by ready-made software-as-a-service or outsourcing. Business demands for faster, better, and cheaper delivery are driving new development and operations methodologies. And demands to address disruption in the marketplace are driving increased innovation and experimentation with a variety of new vendors and other partners. In short, IT is facing a "double-whammy" of disruption—in the business it is part of and in the IT industry itself. Managing this dual disruption "is a bit like playing whack-a-mole," said one IT manager. "You just get one issue under control and two or three more pop their heads up!"

Disruption in IT is not a new problem, said the focus group, noting that new technologies and modes of delivery have long been part of the ongoing evolution of their functions. However, they did agree that moving to new modes of infrastructure and service delivery is likely to have dramatic impacts on the structure and function of the IT organization of the future. As in the business, disruption in IT is about more than new technology (Kushida et al. 2015). In both areas, successful disruption solves problems, some of which are obvious and others that customers didn't know they had until they were presented with an alternative (Adams 2016). The most obvious pain point between IT and the business has long been cost. Cloud computing and outsourcing both reduce the cost of technology by enabling it to be shared. While this is not a new idea, the combination of vastly improved connectivity and new operational software has given cloud computing "legs." As a result, "IT now has a mandate to choose software-as-a-service wherever possible," said a manager.

Another problem these disruptions address is responsiveness. The focus group admitted that their organizations were not only seen as expensive but "mind-numbingly slow" to process business requests. "It takes our organization many meetings to answer a question posed by the business when all they want is an opinion," said a manager. Many IT processes are both slow and cumbersome and not designed to function in a business environment requiring both speed and flexibility. "There is much frustration both in IT and in the business with the time IT takes when we need to be nimble," said another manager. "Our IT is currently structured for efficiency, not effectiveness," said a third. "We need to change how we work and this may cost more."

IT processes are such a significant obstacle to change that some organizations may just give up the effort. In one case, IT's unwillingness to change led to the wholesale outsourcing of its operations function. "This demonstrates that if we get hidebound in IT, disruption will be wrought upon us," said a manager. Changing internal processes is a challenge, given IT's inclination for making things stable and structured. Often senior IT management is either not aware of the problems caused or is unwilling to make the effort to oversee true and extensive process change. "In many cases, we could fix the process but it's easier to outsource it and let them solve the problems," admitted a manager.

The focus group was adamant that disruption in IT should be seen as an opportunity and healthy change rather than a threat to its existence. "Much disruption is coming from our business' increased appetites for digital capabilities," said a manager. "They are now willing to pay for us to develop or acquire these capabilities in the best way we can." "Disruption can be a problem if it's just you," said another. "But if you work with partners, you can bring in the best ideas." This positive attitude and can-do spirit does not underestimate the scope of the changes that will be needed to proactively and successfully manage the significant disruptions that are currently underway, to a greater or lesser degree, in all IT functions.

Disruption and the Future of IT

"Cloud is forcing everyone to rethink IT," said a manager. In business, less than a decade ago a new company would have to invest substantial funds and time in developing IT systems so it could compete. Today these are available on a pay-as-you-go basis from anywhere in the world at a much lower cost (Griffin 2015). The good and bad news is that cloud lowers the bar for accessing both commodity IT services and innovative ones (Kushida et al. 2015). From an IT point of view, cloud makes it easier to deploy standardized applications (e.g., for HR, CRM) quickly and across organizational silos. It also enables rapid experimentation and prototyping at a relatively low cost. And it dramatically changes the "how" of computing, "driving a fundamental paradigm shift from the computing of scarcity to the computing of abundance" (Kushida et al. 2015). IT organizations are trying to adapt accordingly by using cloud services wherever possible and saving scarce internal resources for mission-critical IT activities.

More complex services and larger outsourcing relationships are also evolving to address disruption. However, instead of rigid contracts based on uptime and break/fix, companies are looking for more flexible and shorter-term contracts based on a business solution purchase that are more fluid in pricing, structure, and terms (IDG 2016). Since it is rare that a single provider will offer all the components an organization needs, the job of IT will increasingly involve integrating several agreements and determining the optimal combination of providers and skills (IDG 2016).

The focus group agreed with these predictions. "IT will eventually become a portfolio of services provided by arm's length providers," said a manager. As a result, tight central control and delivery of technology will diminish and IT applications will increasingly become a business-managed concern. Already, the business is becoming directly involved in developing applications through agile methodologies, and digital demands are outpacing foundational IT work. "IT will increasingly focus less on the digital layer in-house and seek to achieve speed through developing partnerships," said another.

The focus group believes that IT will therefore become more of a planning and integration function in the future, dealing with the many new variables, risks, and uncertainties introduced by the organization's new partnerships and delivery mechanisms (Weldon and Rowsell-Jones 2015). Key differentiating activities will be kept in-house, while commoditized services, such as much of operations, will be outsourced. "But we can't outsource risk management," explained a manager, "so this will still be an important IT function."

There are still a number of roadblocks to achieving this ideal future state. These fall into three major categories:

1. ***Processes and structures.*** IT's existing processes can inhibit what is possible to do and how people think about these possibilities. The group identified funding processes and governance as being significant potential roadblocks, as well as line-of-business thinking. "We need to shift to enterprise and design thinking and focus more on the overall ease of doing business with our company," said one member. This has been shown to be a major contributor to the success of a company (Adams 2016). More emphasis should also be placed on organizing IT to support value chains rather than on achieving efficiency. "We must stress IT services, not salaries; and business value, not cost. Our business now

ranks agility and capabilities higher than cost," said a manager.

2. *People.* A new type of IT function will call for new IT capabilities and skills. "Many of our people can't change with a process and don't have the skills to undertake new IT roles," said one manager. Nevertheless, the focus group was optimistic. "I'm always surprised at how positive the outcomes of a change are," another said. "People see change as an opportunity and want to learn." Clearly efforts need to be made to retrain existing staff, while also bringing in new people to catalyze change. Many companies are also successfully using innovation hubs to expose their staff to new ways of thinking and working (Smith and McKeen 2016).

3. *Technology.* Almost all existing companies are encumbered with existing legacy systems. These are often based on brittle infrastructure and inhibit changes in the ways that information is processed, stored, and used (Kushida et al. 2016). It is a reality that these systems cannot always be replaced by software-as-a-service as yet and so must be supported and operated. "We've had a directive to use software-as-a-service wherever possible for the past five years, but we sometimes need to build our own applications," said a manager. "And we still have over 1,000 applications to support, compared with about 150 cloud applications."

Managing the evolution toward the IT organization of the future involves addressing each of these areas as well as understanding and coordinating the interplay between new technology use, disruption, and innovation, and individuals, stable processes, emerging processes, and, more broadly, organizational structure, strategy, and performance (Yeo and Marquardt 2015).

Managing Disruption in IT

Despite fears that disruptive influences are lurking just around the corner, there is widespread agreement that disruption can be managed. "Innovation and disruption are linked but are not the same," explained a manager. "Innovation is a *proactive* response to disruption. Our job is to continually look for new opportunities." Most organizations today are seeking ways to proactively incorporate disruption *before* it is forced upon them and there are several ways to do this effectively (Sheffi 2015; Gans 2016; Adams 2016). These general approaches to managing disruption must then be adapted to reflect specific contexts and functions.

The first goal of managing disruption is to *stay relevant*. And the first step towards this is changing attitudes and behaviors, processes, and structures to make them more flexible and positive towards change. This requires ensuring that an organization's sense of purpose is embedded in its culture and its common outcomes and that this purpose is both compelling for customers and engaging for employees (Yeo and Marquardt 2015). Often those affected by a change tend to focus on the problems involved rather than on the bigger picture. They may focus on how new technology could affect their work rather than its impact on service delivery. These negative views can constrain change, but engaging entire groups to look at their work from a customer's point of view or as a learning opportunity can enable the needed changes (Yeo and Marquardt 2015). Leaders in both IT and the business must provide this overall context so the people involved can work together to make shared decisions and take a positive stance toward change.

To facilitate this goal, senior leaders must be clear about their strategy and values (Beer et al. 2016). If they are not committed to a new direction or do not adopt changes in their own behavior, this will prevent honest conversation about problems that arise, such as poor coordination due to organizational design, talent issues, and barriers to effectiveness in processes. "These barriers almost always appear together and block the systemic changes needed," notes one study (Beer et al. 2016). Effectively managing disruption means being open to redesigning organizational roles, responsibilities, and relationships to overcome barriers, undertaking process consultations, coaching to help people become more effective in a new design, adjusting organizational and individual performance metrics to incorporate new expectations, and providing training where needed. In short, it is essential to recognize that the *primary target* for managing disruption is the *organization*, followed by training and education for individuals (Beer et al. 2016).

The key to being able to manage *specific* disruptions is to be able to detect them quickly and respond appropriately. There is no perfect way to do this and not all disruptions will be detectable (Sheffi 2015). Nevertheless, business and IT leaders must be vigilant together in scanning for both immediate disruptions and potential future ones. Ideally, they should create joint capabilities to explore and examine which disruptive elements are most critical to the organization (Yokelson et al. 2016). This means being able to track disruptions, prioritize their importance, and create practical approaches to dealing with them. Doing this effectively requires incorporating both business and technical contexts (Yeo and Marquardt 2015). It also means taking out *learning options* through experiments and proofs of concept that explore how to make sense of new technologies and ways of working, said the focus group. "Our senior leaders must carve out time to learn and understand the potential of new technologies," said a focus group member. "We don't want disruption for disruption's sake. We want change to be planned and to have an expected outcome."

Finally, organizations must re-evaluate how they approach and manage risk. Traditionally, IT's approach to technology has been risk-averse (Welson and Rowsell-Jones 2015). Now it is being asked to become more agile, experimental, and open to risk in order to deal with disruptive influences. "Our business leaders must recognize that more risk comes with these new approaches," said a focus group manager. "They must also understand that many of our experiments will be designed to learn one thing and fail fast and are not slated for implementation. As well, we will need staff and time to convert successful proofs of concept into a stable reality." Organizations still need reliable, secure, and cost-effective technologies to run their ongoing business. New and traditional approaches to IT work require different processes of funding and governance and different success metrics. Ultimately, however, both must integrate into an operating environment. Recognizing that bimodal (or two-speed) IT is essential to managing disruption, organizations will also need capabilities to bridge the gaps between the two modes of working (Plummer et al. 2016b).

Driving Delivery with Disruption

With these general principles of managing disruption in mind, the focus group had several IT-specific recommendations for IT managers:

- *Embrace disruption.* "Disruption is here to stay," said a manager. "No organization or industry is immune. There are only different flavors of disruption." The focus group agreed that the worst thing an IT leader can do is ignore disruption. "If you identify a disruption and don't do something about it, it just becomes bigger and bigger," said another. They stressed that planning for disruption is essential. "We thought we had it covered with a new innovation group and methods but we didn't think through our business model and user experience," admitted a manager. "We now understand that we need to think things through differently."

Having a big vision is important, they noted, but it is equally essential to pay attention to the little things as well. "Disruptive change is easy to do badly," said a manager. The focus group described several situations where people were laid off due to new forms of delivery, and significant knowledge was lost. "When displacing people, the goal should be minimal impact. It's important to slice finely," said a manager. When changes are made, leaders should be open to hearing from people about their pain points. "Making sure that little things are working can make all the difference," a manager noted.

The group underscored the value of looking outside the organization for new ideas and of challenging assumptions. "Ideally, we should be bringing people together from many different backgrounds and industries, including customers, vendors, suppliers, and start-up firms, as well as other companies, to undertake experiments and build ecosystems," said a manager. "It's important for IT leaders to ask people outside of IT what it can do differently," said another.

- *Make innovation a priority.* Because innovation is the proactive response to disruption, IT leaders should give careful thought to how to enable it in their organizations. Facilitating innovation requires working at the intersection of what is possible with technology, desirable for customers, and viable in the marketplace. Thus successful innovation involves a number of components both inside and outside the organization and close collaboration between them. It is also important to recognize that innovative concepts need to be comprehensively deployed. This means that both IT and business must commit not only to the additional resources needed to develop and integrate experimental designs and technologies with the organization's production technical environment, but also to the time and resources needed to make the process and structural changes required for their success. This commitment goes far beyond that required to develop a proof of concept but it's fundamental to successful transformation.

• *Redesign enterprise architecture.* The focus group agreed that the enterprise architecture group in IT is the right function to drive efforts to identify and manage disruption. "They can see things the business can't because of their role," said a manager. However, this also means the group must change its role and look at architecture's job differently. "Enterprise Architecture needs to aim to get ahead of the curve and become more evangelical and consultative," added another. Unfortunately, many architects still live in an ivory tower, they said, and don't want to change.

A redesigned architecture function must be able to scan and prioritize new technologies and work with business and IT leaders to explore new modes of delivery, new applications, and new ways of working that combine both business and technical knowledge. It will also advise on the related policy and governance issues of initiatives such as cloud services or new forms of outsourcing. Most importantly, the architects involved must be open to change. As one expert noted, "They must hold strong opinions weakly. The worst mistake is to overly rely on one piece of strong information or an assumption, [but] the second worst mistake is to not form an opinion at all" (Yokelson et al. 2016). Finally, they must always be scanning to determine how much change is too much, keeping an eye on its impact on organizational stability, risk levels, and people, and advising the leaders accordingly.

- *Insist on business participation.* It is wrong to leave managing disruption to the CIO alone, said the focus group. "The best opportunities come from adding the right context to them," said a manager. As many organizations have found when adopting new agile methodologies, insisting on active business participation results in much better outcomes. This is not a new concept, but it gains further urgency when it comes to disruption. "IT can see things that the business cannot see because it is focused too narrowly," said a manager. "But the business adds knowledge of the customer and context. Working together clarifies our vision and helps us establish options for the future." "We can ask the business different questions and help them look in different directions," added another. "This helps them think more broadly about the future."

• *Develop new capabilities.* It goes without saying that IT organizations will need new capabilities to successfully manage disruption. The focus group suggested that IT needs new external competencies that will help them better understand their end customers and how to build and leverage ecosystems. This is consistent with new research about innovation that has found that the most successful innovations are those that help customers get a job done well (Adams 2016). Internally, more attention needs to be paid to process assessment and redesign since many IT processes, such as technology lifecycle management, risk management, and incident management, may not work with new business models or new forms of technology or ways of working.

Conclusion

Disruption in the global economy driven by new technologies is a fundamental shift in production paradigms that will have far broader impacts than how it affects individual IT functions or organizations (Kushida et al. 2015). Nevertheless, it is incumbent on IT leaders to take steps to lead their organizations into this uncertain future not only by demonstrating *what* must be done but also *how* it should be done. Managing disruption in IT is complex. It requires IT to develop or improve its skills in many areas, most especially in its partnership with business. It also involves all levels of the organization, from the top down. In this chapter we have pointed to the need for IT to transform itself in order to be successful in managing disruption. This can best be done by first improving the *IT function* and how it facilitates change. It is also time to take a closer look at the organizational processes, structures, and lines of communication both within IT and between IT and the business that inhibit innovation and speed of change. "You can't say this isn't your job anymore," concluded an IT manager. "This simply must be done."

References

Adams, S. "Clayton Christensen on What He Got Wrong about Disruptive Innovation." *Forbes.com*, October 3, 2016.

Beer, M., M. Finnstrom, and D. Schrader. "Why Leadership Training Fails." *Harvard Business Review* 94, no. 10 (October 2016).

Berman, S., and A. Marshall. "Reinventing the Rules of Engagement: Three Strategies for Winning the Information Technology Race." *Strategy and Leadership* 42, no. 4 (2014): 32–22.

Gans, J. "Keep Calm and Manage Disruption." *MIT Sloan Management Review* 57, no. 3 (Spring 2016): 83–90.

Griffin, M. "CIOs Need to Plan and Prepare for Disruption." *CIO*, June 2, 2015.

Grossman, R. "The Industries that are Being Disrupted the Most by Digital." *Harvard Business Review* digital article, March 21, 2016.

IDG Contributing Editor. "Like Everything Else, IT Service Contracts Ripe for Disruption." *CIO*, September 1, 2016.

Kushida, K., J. Muarray, and J. Zysman. "Cloud Computing: From Scarcity to Abundance." *Journal of Industry, Competition and Trade* 15, no. 1 (March 2015).

Plummer, D., D. Smith, and D. Yockelson (a). "Disruptions and Disruptors are Reshaping the Digital Landscape." Gartner Research Report, ID: G00308591, August 17, 2016.

Plummer, D. et al. (b). "Top Strategic Predictions for 2017 and Beyond: Surviving the Storm Winds of Digital Disruption." Gartner Research Report, ID: G00315910, October 14, 2016.

Sheffi, Y. "Preparing for Disruptions Through Early Detection." *MIT Sloan Management Review* 57, no. 1 (Fall 2015): 31–42.

Smith, D., D. Plummer, and D. Yokelson. "Disruptions and Disruptors: Use Digital Business Lenses to Uncover Secondary Disruptions." Gartner Research Report, ID: G00316474, October 5, 2016.

Smith, H. A., and J. D. McKeen. *IT Strategy: Issues and Practices 3rd. Ed.* Upper Saddle River, NJ: Pearson, 2015.

Smith, H. A., and J. D. McKeen. "Leveraging Your Business Ecosystem." *The CIO Brief* 22, no. 3 (2016).

Weill, P., and S. Woerner. "Thriving in an Increasingly Digital Ecosystem." *MIT Sloan Management Review* 56, no. 4 (Summer 2015): 27–34.

Weldon, L., and A. Rowsell-Jones. "Three Steps to Help CIOs Anticipate and Respond to Digital Disruption." Gartner Research Report, ID: G0020156, April 21, 2015.

Yeo, R., and M. Marquardt. "Think Before You Act: Organizing Structures of Action in Technology-induced Change." *Journal of Organizational Change Management* 28, no. 4 (2015): 511–28.

Yokelson, D., D. Smith, and D. Plummer. "Disruption and Disruptors: Differentiating Disruption from Features." Gartner Research Report, ID: G00316473, October 31, 2016.

Chapter 9

DevOps

Because innovation does not produce value until it is fully functioning, considerable effort has been expended to speed up delivery. This effort has largely focused on shortening the development process. But speeding up development is only half the battle. Delivery also requires the completion of an intricate set of implementation procedures to ensure, for example, that the production environment is ready, conversion is orderly, backup plans exist, the infrastructure adheres to architectural standards, and documentation is complete. These activities are collectively referred to as "operations." We have learned that when development proceeds without adequate input, design, and coordination from downstream operations, delivery can be delayed. Innovation with IT is driven much more rapidly and effectively by bridging development and operations, or what is now commonly referred to as "DevOps."

DevOps, a concatenation of the terms "development" and "operations," is a relatively new development method that some organizations are adopting to speed up the time between an application's preliminary approval and its eventual implementation (Wikipedia 2017). Faster delivery of applications is as challenging for both infrastructure and operations as it is for development and delivery, if not more so. DevOps is the result of the recognition that neither group can go it alone to achieve this but must work together with a common set of processes and tools (Forrester 2015). It is widely agreed that the overall goal of DevOps is to promote collaboration between Development and Operations staff to improve software application release management by standardizing development environments and maximizing the predictability, efficiency, security, and maintainability of operations processes by making them more programmable and dynamic. However, the detailed definition of what DevOps *is*, is still unclear (Barros et al. 2015; Mueller 2015), and implementing it remains controversial because it challenges conventional IT thinking (Colville 2014). Practitioners point out that a successful DevOps requires a fundamental transformation of how the IT function works, and this involves considerable effort since, as one manager explained, "IT is a big ship to turn around."

Although DevOps adoption is growing in companies, one study found that only 16 percent of firms are currently using it for production projects. By far the largest number of organizations surveyed are still watching its development (40 percent) or planning to implement it in the next two years (31 percent) (Head and Spafford 2015). Those using DevOps claim that it improves deployment frequency, which can lead to faster time to market, lower failure rate of new releases, shortened lead time between fixes, and faster mean time to recovery in the event of a new release crashing. However, many remain skeptical, calling DevOps the latest IT "shiny new object" (Colville 2014). Gartner Group predicts that by 2018, at least 50 percent of organizations using DevOps principles will not be delivering the benefits stated in their original business cases (Head et al. 2015).

Transforming IT to incorporate DevOps involves first developing a better understanding of what problems this approach is attempting to solve, and then exploring what exactly DevOps *is* and the value it can deliver. Following this, managers should be aware of the obstacles they could face when attempting to introduce DevOps, and then begin to develop a multi-level plan to implement it successfully.

What Problem Is DevOps Solving?

In recent years, agile methodologies have become increasingly accepted in IT as a way to speed up the creation of new applications. These emphasize collaboration and communication between the analysis, design, and development functions in

IT and business users, and are now widely utilized for applications using new technologies (e.g., mobile, social) that are designed to interface with customers.[1] Although these agile methodologies encourage collaboration in *development*, there is rarely integration of these activities with IT *operations*, and this gap causes significant delays in the delivery of new functionality into actual production where it can be used.

With increasing pressure from business to deliver more applications faster without compromising quality, many companies are now recognizing that operations is a bottleneck. "What's the point of agile development if the output of an iteration makes it only as far as a 'test' or 'staging' environment?," said a focus group manager. "People [have begun] to realize that a 'throw it over the wall' mentality from development to operations is just as much of a challenge as [IT] used to have with throwing [business] requirements over the wall and waiting for code to come back" (Greene 2015). "Traditionally, provisioning has been a lengthy process in operations," said a manager. "Moving into production requires many steps: bundling the code, packaging the product, provisioning the environment, and coordinating with the operations staff. It has not been common to release code into production regularly. But we can't wait sixteen weeks for this to be done anymore."

Focus group members added that there is also a fundamental difference of mindsets between the development and

---

[1] More traditional waterfall methods still tend to hang on in many organizations for larger and more back-end systems.

operations functions, which causes problems in communication and goals. "Development strives for change," said a manager. "With agile, especially, there are new change sets that need to be rolled out to production frequently. But operations strives for stability to make sure all environments run smoothly and without costly service disruptions." This disconnect between development and operations has created a "wall of confusion" between these two areas, that is further exacerbated by IT management structures that often have the two areas reporting separately all the way up to the CIO. As a result, it is not uncommon for the groups to blame each other for unplanned outages and system failures (Edmead 2015).

Finally, there is also recognition that operations itself hasn't changed to keep up with new technological advances. Twenty-first-century operations will increasingly be more virtual than physical, and existing operations principles, processes, and practices have not kept pace with what's needed for success (Wooten 2015). "Resources may be physical or they may just be software—a network port, a disk drive, or a CPU that has nothing to do with a physical entity" (Loukides 2012). With applications running on multiple virtual machines, they need to be configured identically and managed for resilience. Companies can no longer afford system outages due to failures or downtime for upgrades. Instead, operations specialists must learn to design and work with infrastructure that is monitored, managed, and reconfigured all in real time without interrupting service, in the same way a few large, cloud-based service providers have learned to do.

Making operations more effective, reliable, and flexible therefore has several components:

- Breaking down organizational and communications silos both within operations and between operations and development to facilitate improved productivity;

- Re-engineering operations procedures to streamline them and reduce waiting time;

- Designing operational procedures to be more reproducible and programmable and reduce human error;

- Enabling more frequent production releases of code;

- Working with all parts of IT to deliver business outcomes, rather than technical metrics.

This is where the principles and practices of DevOps come in.

What Is DevOps?

To a large extent DevOps is an extension of agile principles beyond the boundaries of development to the entire delivery pipeline, since "software isn't done until it's successfully delivered to a user and meets expectations around availability, performance, and pace of change" (Mueller 2015). Extending these principles to operations means that this function must begin to think in terms of delivering an end-to-end service focused on a business outcome, rather than sub-optimizing on their own particular task or, as is often the case, sub-tasks. This has significant implications for the people, processes, and tools involved in both development and operations.

Defining DevOps is challenging because there is no common set of practices, descriptions, or even shared meaning about this concept (Haight 2015b). Instead, there's such a variety of hype, marketing, and commercialized tools promoting it that practitioners can easily perceive DevOps to be an intimidating transformation of Herculean proportions (Greene 2015). In fact, to some, DevOps is about extending agile principles to *all* IT functions, such as quality assurance, security, architecture, and testing, in addition to operations (Wooten 2015). While this is probably desirable at some point in the future, focusing more specifically on how to bring operations and development together should be the primary initial focus for any organization exploring DevOps, said the focus group.

DevOps can be defined as:

A set of practices, tools, and policies based on agile principles which aim to bring infrastructure setup and awareness earlier into the development cycle and to establish consistent, reliable, and repeatable automated deployments. (Adapted from Greene 2015)

It is based on five key principles (Haight 2015a):

1. *Iterative.* DevOps is similar to agile methodologies in that it facilitates projects with high degrees of uncertainty and does not require comprehensive plans.

2. *Continuous.* The ultimate goal of DevOps is continuous delivery of projects into staging and then deployment. It supports rapid change and is never "done."

3. *Collaborative.* All stakeholders share common objectives, trust each other, work together to solve problems, and agree on the mission and metrics of the project.

4. *Systemic.* DevOps applies agile principles to operations and seeks to remove as many constraints to delivery as possible in this function.

5. *Automated.* Wherever practical, automated tools are adopted to link the different steps of the DevOps process—from development and testing, to integration and deployment—to deliver fast, safe, and cost effective solutions, optimizing the entire value chain.

"The goal is 'No Surprises' on delivery," said a focus group manager. At present, many defects and errors appear only when an application moves from the development environment to the testing or production environments, which are often configured somewhat differently. In DevOps, developers and operations specialists work together to specify configuration standards in the development environment, and these continue to be used in the production environment, thereby removing a significant source of error and eliminating considerable elapsed time. Similarly, developers and operations staff work with common toolsets and a common language to achieve common metrics (Forrester 2015). This prevents miscommunication and creates a culture of collaboration, cooperation, and trust that has a wide range of benefits including:

- *Higher-quality software delivered faster and with better compliance.* DevOps tries to prevent issues during development, rather than resolving them during operations. Removal of "friction" between software delivery steps also improves timelines (Forrester 2015).

- *Rapid resolution of problems* with development and operations working together in cross-functional teams in a "no-blame" environment to fix errors and move fixes quickly into production.

- *Reduced risk.* Having the full range of development and operations skills on a project reduces the need for risky handovers to other groups (Wooten 2015). Continuous implementation means that any new release contains fewer changes, making them easier to identify and fix, or roll out as necessary (Greene 2015).

- *Improved efficiency.* Using DevOps tools to automate the development–operations "conveyor belt" speeds up time to market and removes obstacles to implementation so that teams can iteratively implement new functionality reliably and predictably (Wooten 2015). "We've seen a near doubling of output per developer through automating repeatable tasks, freeing them up to focus on higher value tasks," said one focus group manager.

- *More resilient systems and operations.* Resilient systems are those that can adjust their functions before, during, and after an unexpected event (Webb 2011). Operations people are specialists in resilience and can build monitoring techniques into applications that enable more resilient applications. At the same time, automated operations tools make it easier to roll changes back if problems occur and to deploy new ones (Loukides 2012).

- *Improved customer experience and satisfaction.* In contrast to existing process design approaches (e.g., ITIL, Lean, TQM) that look at an initiative from the organization's perspective, DevOps explicitly incorporates recognition of the customer and customer experience into its initiatives by focusing on issues that really matter to them (e.g., utility, convenience, value, delight) (Haight et al. 2015). It focuses on the work that needs to be done to deliver something that is actually usable by a customer (Denning 2015). Internal customers like the continuous delivery of new functionality to the marketplace and the fact that IT is delivering strong stable functions (Loukides 2012).

Focus group members recognized the *potential* for DevOps to deliver these types of benefits, although none had DevOps fully implemented. "We've coordinated all our provisioning processes and seen infrastructure configurations go from taking six months to 30 seconds," said one manager. "We're working on hardware and software standardization so we can create a release path for applications," said another. "Standards are the key but they're tough to implement because there are so many tools involved." "We are having the right conversations and the ideas are percolating through our organization," said a third. "We're more aware of the possibilities and would like to increase the speed of automation between development and operations," stated a fourth.

At present it is clear that DevOps remains more a concept than a reality in most companies. However, as agile principles take root and with continuous pressure to deliver more, better, and faster from business, IT leaders are now taking very serious looks at how DevOps can help them achieve improved business outcomes faster and more effectively. There was a general recognition in the focus group that some or all DevOps concepts are the right way to move forward to improve IT delivery and resiliency. Some have begun pilots within their "new technology" areas to experiment with these ideas and tools. It is just going to be very challenging to make these changes. "Getting DevOps started in Development was relatively easy but integrating it into Operations is a much larger challenge," concluded a focus group manager.

DevOps Challenges

The vagueness of the DevOps definition and the lack of shared understanding about its practices have prevented many organizations from implementing it or implementing it well (Colville 2014; Wooten 2015; Haight 2015B). The focus group managers recognized that there are many other challenges to be addressed by IT in order to ensure the successful implementation of DevOps, including:

- *Current organizational structures.* DevOps requires the development of cross-functional teams with "full stack" understanding of code, systems administration, and infrastructure that are not the norm today (Wooten 2015). Many IT organization structures inhibit collaboration due to their siloed design, and this is one of the most important obstacles for IT managers to consider when thinking about implementing DevOps. "If you want DevOps, you must change how people are aligned," explained a focus group manager. Operations is still seen as a separate function with its own skills and mindset. Deploying DevOps therefore requires breaking down the organizational and cultural walls that have been built between the groups. Even within operations, there are often specialized sub-groups that focus on specific areas of the infrastructure, such as storage or networks.

- *Lack of management support.* As with agile development, business wants the benefits of DevOps but doesn't always support it (Head et al. 2015). IT leaders struggle both with understanding it themselves and also with explaining it to the business. "Transforming to DevOps requires a complete reorganization for end-to-end service," explained a manager. "To make DevOps work we need management to support this with vision, clear expectations, proper tools and software, integration with existing processes and the legacy environment, mindset changes, and general awareness and communication of what is involved," said another. "Our bureaucracy is very slow to change."

- *Organizational constraints.* Other IT groups can also place limitations on how DevOps is implemented in an organization. Architecture, compliance, security, quality assurance, and current legacy environments can each inhibit what can be done and must be considered when adopting this method (Head et al. 2015). In particular, security functions can be concerned about the fact that production code is in a constant state of change and that vulnerabilities could be introduced at any time. Architecture can play an important role in DevOps if it chooses, said the focus group. It can help the organization develop standards for data, integration, and infrastructure, which can facilitate its adoption, as well as recognize repeatable patterns that can be automated and identify appropriate tools for these processes. In operations, ITIL and change management practices developed over many years can be in direct contradiction with DevOps principles, which makes some organizations queasy. "There are reasons for some of these walls that have to do with segregation of duty," said a manager. "We need to be careful about how we break them down."

- *Lack of budgets.* A transformation to DevOps requires funding to train existing staff and hire experienced outside staff. Tight budgets and lack of time for training can leave people lacking skills and even understanding of the goals of DevOps (Forrester 2015). "In a project-driven environment, the question is always, 'Who pays?'" said one manager. "We need some elasticity in our budgets to make these changes." Pressures to keep staff lean are particularly strong in operations and mean that it is difficult to get staff from this group to participate in DevOps projects, said the focus group. "Operations is usually too busy for DevOps," commented one manager. "They can't focus on helping developers."

- *Resistance to change.* "Culture is a huge factor in changing to DevOps," said a manager. "There is real tension between the old and new ways of working," said another. "People end up gaming the system." A number of individual factors contribute to resistance to DevOps. First, it can be intimidating, requiring IT professionals to understand and use a wide variety of new tools and processes that are difficult to understand (Greene 2015). Second, DevOps changes project methods, operational practices, culture, and tools (Head and Spafford 2015). In particular, new tools automate parts of operations workflow and need operations staff to develop skills working with code-like languages, with which they are not comfortable or familiar. This can lead to fear of loss of control or even loss of their jobs, said focus group members. "We must be very sensitive about this and communicate clearly about these changes," a manager said. Third, neither individuals nor organizations are accustomed to the pace of change involved in DevOps (Head et al. 2015). For operations, the concept of constant releases into production, rather than staged periodic releases, as has been traditional, requires a total change of mindset, and this generates both questions and resistance (Wooten 2015).

Thus there is widespread recognition that culture plays a significant role in both enabling and inhibiting DevOps success. This perspective was widely recognized in the focus group discussion. "There must be proactive education about how to take on the risks involved in DevOps," said a manager. Another manager noted, "We know we have to get our act together or we will be 'clouded.'"

- *Tools.* Automating configuration and the transitions involved in moving from development to testing to production environments is a completely new way of working—both for development teams and for operations specialists—and involves several sets of new tools. DevOps experts speak about "tool chains" where teams create, build, test, version, and deploy a program with outputs from one or more tools acting as inputs into other tools in a pipeline fashion (Haight 2011). Naturally, there has been a literal explosion of new tools to work in this space (Mueller 2015), and many technologists gravitate towards these when deploying DevOps (Wooten 2015). However, DevOps writers stress that DevOps is not about tools but about culture (Loukides 2012), and the focus group agreed. The group also noted that tool vendors are putting increasing pressure on IT to change by continuing to approach business leaders with their new DevOps tools. "It's a constant struggle to stay ahead of the business in this area," said one manager. "We need to find the appropriate tools for us."

- *Collaboration difficulties.* As noted above, operations and development have different mindsets, speak in different languages, and use different tools and metrics. This can result in miscommunication, misunderstanding, and errors. Also, the groups have competing agendas. Development wants to release and test more often and spend less time waiting for someone to configure applications and infrastructure. Operations wants stability and efficiency. As a result, a survey of companies using DevOps found that collaboration difficulties between operations and development and lack of common skills were their top two challenges (Forrester 2015).

- ***The need to integrate service providers.*** Most organizations deal with an ecosystem of different service providers for a variety of services. Where providers deliver some parts of IT functionality, the DevOps challenge becomes even trickier. Organizations wishing to adopt DevOps must first verify that their service providers are able to implement them (Barros et al. 2015). At present very few service agreements (10 percent) require DevOps delivery (Overby 2015), so IT organizations working with service providers will need to address how to adapt their agreements and expectations with them in order to benefit from this new trend.

- ***Lack of skills.*** Transforming to DevOps requires the development of skills and capabilities in IT that involve new ways of working, new common languages, new processes, and the ability to work with new tools. Operations specialists on teams are now expected to write the code that will maintain the organization's infrastructure and ensure application resilience (Loukides 2012). New skills in this area include: automation engineering, process ownership, and product management (Eliott 2015). Often these skills are not available in-house and must be hired from outside. Poor process disciplines can also undermine DevOps, especially when more traditional methods are still the dominant portion of IT work (Questex 2015). One study found that both development and operations are skeptical about whether the *other* group has the time or motivation to learn new skills (Forrester 2015). Another study found that lack of investment in developing DevOps skills is a key warning sign of the potential failure of this implementation (Head et al. 2015). Finally, IT leadership must also develop new skills to overlay the right amount of governance and decision-making on DevOps teams (Cana 2015).

• *Inadequate fundamentals available.* Such a dramatic transformation in IT requires more than just skills and tools, said the focus group. There are a number of fundamentals that must be put in place for DevOps to be successful. Chief among these are standards, including configuration standards, data standards, and integration standards. Complexity in each of these areas works against what can be done with DevOps, they said, and inhibits the pace of automation. The proliferation of tools also makes standardization tough since they don't always work in a "plug 'n play" fashion, the group said.

Clearly, although desirable in the longer term, DevOps is a huge transformative challenge for IT organizations. "Times are changing and we have to change with them," said a manager, "but people are not always willing to adapt to the new way of doing things. We have to expect some tension along the way."

Driving Delivery with DevOps

The focus group stressed that there is no single recipe for implementing DevOps. "This is a problem we're trying to solve," said a manager. "For a while it will be an evolution, with many different models of application development running across the full spectrum of services," said another. Many companies are still struggling to adopt agile principles and methods, so the extension of these into DevOps can appear overwhelming. Since the IT industry is still at the early stages of understanding what DevOps *is*, taking the next step of implementing it can be especially difficult. Nevertheless, DevOps is the next obvious step following agile development and the value proposition is compelling. Therefore, it is important for IT managers to at least begin to explore how they might implement DevOps in their own organizations. The focus group had several recommendations about how to do this:

- *Communicate and educate.* The first step to a successful implementation of DevOps, said the focus group, is to define what it means for your organization. Once this has been clarified, the vision, goals, and plan must be communicated to every area of IT from the top down. "Everyone needs to understand the benefit and why processes and structures need to change," said a manager. As the ideas begin to percolate throughout IT, leaders can then begin to establish new expectations around how work gets done.

- *Consult.* Both Dev and Ops staffs should be involved in designing any changes. "Ideally, they should help build the roadmap for achieving full DevOps," said a manager. This process can also identify early the existing obstacles or constraints that will need to be removed. "The goal is to ensure that the right way is the easy way," said a manager.

- *Take incremental steps.* Full DevOps implementation may be some time away for many organizations but there are several ways to "move the ball forward" towards this vision. For example, one focus group organization has now incorporated an operations sign-off for solution approval. Another has started with pilot projects led by experienced staff. "Then, we replace the leads, rinse, and repeat," said the manager, who hopes this strategy will spread DevOps awareness and capabilities throughout IT. Pilots also help demonstrate the value of this approach. Wherever possible, managers should select staff who are the right cultural fit for DevOps and who are open to change. And even though DevOps may be intrinsically more aligned with agile development methods, it is also possible to use some DevOps principles in more traditional development projects to help them improve their consistency and reduce cost (Head and Spafford 2015).

• *Develop cross-functional teams.* A first step in this area is to break down functional silos within operations. "We are creating service streams that focus on the different value components of operations, such as operational readiness, asset management, and performance management, instead of individual technological elements," said a manager. Eliminating operations "towers" and centralizing them promotes more cohesion within operations and is a step toward enabling an operations specialist to represent all operations functions on a DevOps team. Following this, IT teams can be reorganized to deliver end-to-end service with Dev and Ops staff co-located and dedicated to a set of deliverables. "Co-location is key to developing shared goals, common language, and improved communications," said a manager. It also helps to create more resilient systems, ensuring that developers are not divorced from the consequences of their work and that they work together with operations to resolve problems (Loukides 2012).

- ***Establish foundational elements.*** Tools to automate parts of the DevOps delivery process are important but it is also essential to choose ones that work for a particular organization and that work well together, said the group. The full tool chain doesn't need to be implemented at once. Instead, the group recommended focusing on using automation to fill gaps and address pain points that constrain delivery. Standards are especially important foundational elements because they can help an IT organization change in a safe way, eliminate complexity, and help technologies and skills converge successfully. Standard environments can also be introduced for legacy systems. Finally, an important element of ensuring that both Dev and Ops staff are aligned for delivery is to measure them in the same way. Management can therefore promote greater collaboration by adapting current metrics to measure joint Dev and Ops success in achieving business enablement, improving agility, and correcting problems.

- ***Allocate time and money.*** This type of fundamental transformative change takes time and costs money, cautioned the group. If necessary, hire some experienced staff or bring in consultants to lead the first few projects and transfer their knowledge to others. However, wherever possible, the focus group felt that existing staff should be trained to develop DevOps competencies.

- **Coach.** Promoting such a significant transformation requires management to facilitate learning, accept more risk, and stimulate communication and collaboration for improved delivery. This necessitates a different style of IT management that is more focused on coaching and removing obstacles and less on command and control, thereby freeing up DevOps teams to do whatever it takes to deliver quality products for the business.

- **Don't get carried away.** DevOps will take hold at varied rates, said an experienced manager. Expect to see different levels of maturity as these new concepts are introduced. Similarly, not all service providers will be ready to implement them so it's important to assess their readiness for this concept. In addition, there may be some components of DevOps for which an organization may not be ready or areas in which it is not required. "DevOps is a capability that must be used where appropriate, not implemented comprehensively," concluded the focus group.

Conclusion

Implementing DevOps successfully is not for the faint of heart. It involves tackling some fundamental assumptions about how IT is supposed to function—breaking down the structural, procedural, and cultural walls that have traditionally divided the development and operations groups. If done well, it will result in faster time to deliver business value and more reliable systems. However, as the long list of challenges in this chapter illustrates, achieving these benefits will be neither easy nor cheap. Organizations will have to decide for themselves the best time to introduce DevOps. Although it is very tempting to wait until the concept, methods, and tools are better defined, the risk of doing nothing is high. With IT functionality and services changing at an exponential rate, managers will have to balance the challenges of change with the risk of getting blindsided and disintermediated by new cloud-based service providers that can deliver faster, better, and cheaper than IT functions can do with their traditional methods.

References

Barros, D., J. Longwood, and G. van der Heiden. "DevOps Applied in Outsourced Environment Delivers Business Agility." Gartner Research Report, ID: G00272680, April 29, 2015.

Cana, M., and N. Osmond. "How to Get Started and Deploy Bimodal Capabilities in CSPs." Gartner Research Report, ID: G00275202, July 16, 2015.

Colville, R. "Seven Steps to Start your DevOps Initiative." Gartner Research Report, ID: G00270249, September 16, 2014.

Denning, S. "Agile: It's Time to Put it to Use to Manage Business Complexity." *Strategy and Leadership* 43, no. 5 (2015): 10-14.

Edmead, M. "Lean IT and DevOps: The New Kids on the Block." *CIO*, July 21, 2015.

Elliot, S. "What Makes a DevOps Unicorn?" *CIO*, September 9, 2015.

Forrester Research. "Infrastructure as Code: Fueling the Fire for Faster Application Delivery." Forrester Research Inc. https://go.forrester.com/, March 2015.

Greene, D. "What is DevOps?" Crunch Network. http://techcrunch.com/2015/05/15/what-is-devops/#.d3ik6im:5XYN, May 15, 2015.

Haight, C. "Principles and Practices of DevOps." Gartner Research Report, ID: G00272990, March 12, 2015a.

Haight, C. "How to Scale DevOps Beyond the Pilot Stage." Gartner Research Report, ID: G00272992, March 11, 2015b.

Head, C., G. Spafford, and T. Bandopadhyay. "Avoid DevOps Disappointment by Setting Expectations and Taking a Product Approach." Gartner Research Report, ID: G00280974, October 6, 2015.

Head, I., and G. Spafford. "Step 1 in Delivering an Agile I&O Culture is to Know your Target State." Gartner Research Report, ID: G00272648, March 31, 2015.

Loukides, M. "What is DevOps?" *O'Reilly Radar*. http://radar .oreilly.com/print?print_bc=48235, June 7, 2015.

Mueller, E. "What is DevOps?" The Agile Admin, https:// theagileadmin.com/what-is-devops/, November 10, 2015.

Overby, S. "IT Customers Slow to Embrace Outsourced DevOps." *CIO*, August 7, 2015.

Questex Media. "DevOps Will Evolve from a Niche to a Mainstream Strategy, says Gartner." *Networks Asia*, Questex Media Group, March 12, 2015.

Webb, J. "How Resilience Engineering Applies to the Web World." *O'Reilly Radar*. http://radar.oreilly.com/ print?print_bc=46361, May 13, 2011.

Wikipedia. "Devops." https://en.wikipedia.org/wiki/DevOps, accessed October 30, 2017.

Wooten, B. "What is DevOps Not?" http://www.infoq.com/ articles/devops-in-banking, November 23, 2015.

Index

www.ingramcontent.com/pod-product-compliance
Lightning Source LLC
Chambersburg PA
CBHW061150220326
41599CB00025B/4433